A Primer of LISREL

Barbara M. Byrne

A Primer of LISREL

Basic Applications and Programming
for Confirmatory Factor Analytic Models

With 25 Illustrations and 60 Tables

Springer-Verlag
New York Berlin Heidelberg
London Paris Tokyo Hong Kong

BARBARA M. BYRNE, PH.D.
University of Ottawa
School of Psychology
145 Jean-Jacques Lussier
Ottawa, Ontario, Canada K1N 6N5

Library of Congress Cataloging-in-Publication Data
Byrne, Barbara M.
  A primer of LISREL: basic applications and programming for
confirmatory factor analytic models/Barbara M. Byrne.
    p. cm.
  Bibliography: p.
  Includes index.
  ISBN-13:978-1-4613-8887-6
  1. LISREL (Computer program)   2. Social sciences—Statistical
methods.   3. Analysis of variance.   I. Title.
HA32.B97   1989
519.5′354′028553—dc20   89-19657

Printed on acid-free paper

Typeset by David E. Seham Associates, Inc., Metuchen, New Jersey.

9 8 7 6 5 4 3 2 1

ISBN-13:978-1-4613-8887-6   e-ISBN-13:978-1-4613-8885-2
DOI: 10.1007/978-1-4613-8885-2

Dedicated
to Alex for his continued patience, support, and understanding
and
to Mom and Dad for instilling in me the virtues of hard work

Preface

The purpose of this book is to provide a nonmathematical introduction to the LISREL computer program (Jöreskog & Sörbom, 1985). It was written with a very specific audience in mind—those wishing to use the LISREL program, but who, so far, have been frightened off by its seemingly complex notation, mathematical concepts, and overall statistical headiness. Since I too am a victim of math anxiety, symbol shock, and other related phobias, I know only too well, the frustration of trying to translate mathematical jargon into plain and simple everyday language. The book is therefore written not for the mathematically sophisticated, but rather, for those who seek a more earthly approach to the topic.

After some eight years of finding my way through the LISREL forest that included the previous program versions IV and V, and after asking innumerable questions regarding its use, I wish now to share what knowledge I have on applications of LISREL with my fellow math phobics. As with other life experiences, some are best learned by doing: nowhere is this more true than in learning to use LISREL. By providing concrete applications, along with accompanying data and important caveats regarding the LISREL program, I hope to save you from the trial-and-error approach that characterized much of my own learning of the program.

The focus of the book is not on the theoretical or statistical framework of LISREL. Indeed, there are now many well-written texts on the topic of LISREL modeling in general, and many excellent journal articles on particular problems related to it. Rather, the emphasis is on the practical aspect of LISREL modeling. As such, the new user is "walked through" a variety of LISREL applications; all are based on the confirmatory factor analytic model.

All application examples in the book are taken from my own research in the area of self-concept. I chose this approach for several reasons. First, it maximized my freedom to make the data available to you for your own experimentation with the program. I urge you to work through each application using these data. Second, it enabled me to provide you with specific journal references to which you may turn should you wish further elaboration of the underlying theory, measuring instruments, sample description, and the like. Finally, it provided me with detailed information regarding problems encountered at various stages of the data analyses. The value here comes from learning how to solve the problem; I pass this information along to you with each selected application. While these applications derive from a psychological perspective, it should be emphasized that all are equally applicable to any of the other social or behavioral sciences.

The book is divided into three major sections. In Section I, I introduce the reader/user first to basic concepts associated with the LISREL model (Chapter 1), and then to the basic components of the LISREL program (Chapter 2). Section II focuses on single-group analyses; I present applications related to the validation of a theoretical construct (Chapter 3), a measuring instrument (Chapter 4), and multiple constructs assessed by multiple measures within the framework of the multitrait-multimethod matrix (Chapter 5). Finally, in Section III, I examine applications that relate to multigroup analyses. Specifically, I demonstrate procedures for testing the measurement and structural invariance of a theoretical construct (Chapter 6), a measuring instrument (Chapter 7), and latent mean structures (Chapter 8).

With each application, I provide data in correlation matrix form, along with means and standard deviations; a schematic presentation of the model being tested; the specific LISREL program input; a discussion of results in terms of goodness-of-fit and post hoc procedures; and the related journal reference where the theoretical framework, methodology, and substantive findings are described and discussed in more detail. Elsewhere, I present and interpret selected portions of the LISREL output, point out the causes of and solutions to particular error messages, and offer helpful caveats related to particular LISREL functions.

Acknowledgments. I wish to express my indebtedness to several people who are directly responsible for my persistent interest in the application of LISREL to various models of psychological phenomena. To Richard J. Shavelson, without whose continued encouragement over nearly a decade that spans the early days of my doctoral research up to the present, the contents of this book would never have come into being; to Rich I owe much more than I can ever hope to repay. Although initially he knew me only as a signature on a letter requesting more information on a proposed structural model to validate self-concept (Shavelson & Stuart, 1981), he took the time to write and encourage me in my doctoral work, convincing me that my research represented a worthy contribution to the field. More recently, as a post-doctoral fellow at UCLA, where I had the good fortune to collaborate with him on further validation of self-concept using LISREL applications, he taught me other important skills essential to good research: clear thinking, succinct writing, and thorough analyses. In my view, Rich Shavelson is the epitome of both teacher and researcher par excellence; he will forever be an inspiration to me in my academic endeavors!

To Bengt Muthén who, during my tenure at UCLA (and many times since), has answered innumerable questions, provided invaluable advice, and taught me the importance of knowing my data well; his patience, I'm certain, has been tried on many occasions when explanations have required translation into non-mathematical terms that I could understand. To David Kaplan who, during my time at UCLA, helped me resolve a multiplicity of LISREL anomalies, and with whom I enjoyed sparring about the goodness-of-fit of LISREL models based on "real" versus "unreal" data. To Herbert W. Marsh whose plethora of self-concept papers, many of which have involved LISREL applications, have constantly challenged me to learn more. To James E. Carlson, who during the period of my doctoral research, guided me through my first LISREL experiences back in the days of LISREL IV when finding "start values" was akin to looking for a needle in a haystack. To Lee Wolfle, who provided me with the first explanation of LISREL symbols and matrices that I could actually understand; his 1981 AERA paper will forever remain a prized possession in my collection of LISREL papers. And finally, to Peter Bentler whose papers and oral presentations have provided me with a wealth of invaluable information; I can only hope to aspire to his consummate interpretative and literary skills.

BARBARA M. BYRNE

Contents

Section I Introduction

1
The LISREL Confirmatory Factor Analytic (CFA) Model

1. Basic Concepts

1.1. The Role of Latent and Observed Variables

In the social and behavioral sciences, researchers are often interested in studying theoretical constructs that cannot be observed directly; such phenomena are termed *latent variables,* or *factors.* (Throughout the book, the terms "construct," "latent variable," and "factor" are used interchangeably). Examples of latent variables in psychology are self-concept and anxiety; in sociology, powerlessness and racial prejudice; in education, teacher expectancy and verbal ability; in economics, economic expectation and social class.

Since latent variables are not directly observed, they cannot be directly measured. Thus, the researcher must operationally define the latent variable of interest in terms of behavior believed to represent it. Assessment of the construct, then, is obtained indirectly through measurement of some observed behavior. The term "behavior" is used here in its broadest sense to include scores on a particular measuring instrument. Thus, observations may include, for example, self-report responses to an attitude scale, scores on an achievement test, in vivo observation scores related to some physical task or activity, coded response to interview questions, and the like. These measured scores (i.e., measurements) are termed *observed, manifest,* or *indicator* variables; they are considered to represent the underlying construct of interest.

It is now easy to see why methodologists urge researchers to be cautious in their selection of measuring instruments, and emphasize choosing only those that are psychometrically sound. To do otherwise runs the risk of limiting the credibility of the findings.

1.2. The Role of Factor Analysis

The most well-known statistical procedure for investigating relations between a set of observed variables and its underlying constructs is that of *factor analysis.* In using this approach to data analyses, the researcher studies the covariation among the observed variables in order to gather information on a (usually) smaller number of latent variables.

There are two basic types of factor analyses: exploratory factor analysis (EFA) and confirmatory factor analysis (CFA). In EFA, the researcher does hot know the underlying latent variable structure. Thus, the focus of the investigation is directed toward uncovering the minimal number of factors that underlie the observed variables. In CFA, on the other hand, the researcher has some knowledge of the underlying latent variable structure. This knowledge may be based on theory, empirical research, or some combination of both. For example, suppose a measuring instrument is designed to measure four facets of self-concept (say, general, academic, physical, and social), and this factor structure has been validated in the literature. The researcher can feel confident in proceeding with a CFA analysis. As such, he or she postulates a priori that certain test items will be highly related to the latent variables they are designed to measure, and only negligibly related (or, better still, not related at all) to the remaining factors. In factor analysis, these relations are termed *factor loadings.* Thus we say that the items will load highly on those factors for which they were designed to measure and will load negligibly on the other factors. Putting this in context with our example of self-concept, the researcher would specify a priori that the items designed to measure general self-concept would load highly on that factor but would yield loadings of approximately zero on the academic, physical, and social self-concept factors.

While EFA can be conducted using LISREL, it is most commonly conducted using a more traditional approach that can be accomplished by using other statistical packages, such as SPSS, SPSSX, SAS, and BMDP. On the other hand, CFA requires the analysis of covariance structures that is the basis of the LISREL approach to data analysis.[1] This book therefore limits itself to applications that fall within the CFA framework.

1.3. The Role of Statistical Models

Statistical models are a convenient way of describing the structure underlying a set of observed variables. In other words, they provide the simplest explanation of how the observed and latent variables are related to one another. Most people think of statistical models as being geometric schema portraying specific phenomena under study, but this is not always the case; indeed, such diagrams are a very convenient and effective way of getting the idea across. However, statistical models can also be de-

---

[1]Other available computer programs designed for the analysis of covariance structures are EQS (Bentler, 1985) and COSAN (McDonald, 1978) for use with interval data, and LISCOMP (Muthén, 1987) and PRELIS (Jöreskog & Sorböm, 1986) for use with categorical data.

scribed by means of mathematical equations that can be expressed either in matrix or regression format; the latter is a series of regression equations, each of which represents the relation between one observed variable and its underlying latent variable.

Typically, a researcher postulates a statistical model based on his or her knowledge of the related theory, on findings from other research conducted in the area, or on some combination of both. The researcher then sets out to test the model (i.e., test the hypothesis that the model is plausible) by collecting data on all variables specified in the model. The primary statistical problem in this model-testing procedure is to examine the goodness-of-fit between the hypothesized model and the sample data that comprise the observed measurements. Said another way, the researcher imposes structure on the sample data by forcing them to fit the hypothesized model and then determines how well the observed data fit the model under study. Since it is highly unlikely that a perfect fit will exist between the observed data and the hypothesized model, there will be a differential between the two; this differential is termed the *residual*.

The model-fitting process can be summarized as follows:

$$\text{Data} = \text{Model} + \text{Residual}$$

where

*Data represent the observed measurements based on the sample.
*Model represents the hypothesized structure underlying the observed variables.
*Residual represents the difference between the hypothesized model and the observed data.

The statistical theory related to this model-fitting process can be found in the many texts and journal articles devoted to the topic of LISREL modeling.

2. The General LISREL Model

In order to have a comprehensive understanding of the CFA model, it behooves us at this point to first examine the general LISREL model. This diversion is necessary for two reasons: it will help you to more fully comprehend how the CFA model fits into the general LISREL scheme of things and it will facilitate the later understanding of analyses related to mean structures discussed in Section 3.

2.1. Basic Composition

The general LISREL model can be decomposed into two submodels: a measurement model and a structural model. The *measurement model* defines relations between the observed and unobserved variables. In other words, it provides the link between scores on the measuring instruments (i.e., observed indicator variables), and the underlying constructs they are designed to measure (i.e., the unobserved latent variables). The measurement model, then, specifies the pattern by which each measure loads onto a particular factor. The *structural* model defines relations among the unobserved variables. In other words, it specifies which latent variable(s) directly or indirectly influences (i.e., "causes") changes in the values of other latent variables in the model.

One necessary requirement in working with LISREL is that, in specifying the structural model, the researcher distinguishes between latent variables that are exogenous and those that are endogenous. Exogenous latent variables are synonymous with independent variables; they cause fluctuations in the values of other latent variables in the model. Fluctuation in the values of exogenous variables is not explained by the model; rather, they are considered to be influenced by other factors external to the model. Background variables such as sex, age, and socioeconomic status are examples of such external factors. Endogenous variables are synonymous with dependent variables; they are influenced (i.e., "caused") by the exogenous variables in the model, either directly or indirectly. Fluctuation in the values of endogenous variables is said to be explained by the model since all latent variables that influence them are included in the model specification.[2]

2.2. The Link Between Greek and LISREL

In the Joreskog tradition, LISREL models are couched in matrix notation that is represented by Greek letters. Thus, a second necessary requirement in learning to work with LISREL, is to become thoroughly familiar with the various LISREL matrices and the Greek letters that represent them.

In general, matrices are represented by upper-case Greek letters. The elements of these matrices are indicated by lower-case Greek letters; they represent the parameters in the model. By convention, observed measures are represented by Roman letters. As such, exogenous variables are termed "X-variables"; endogenous variables are termed "Y-variables." At the most, eight matrices and four vectors define a general LISREL model.[3] All matrices and vectors, however, may not necessarily be required; this will depend on the particular model specified.

The measurement model is defined by four matrices and one vector; the structural model by four matrices and three vectors. As such, the

---

[2]Although beyond the scope of this volume, it should be noted that in more complex general LISREL models, it is often the case that latent variables operate as exogenous and endogenous variables within the same model.

[3]A *matrix* represents a series of numbers written in rows and columns; each number in the matrix is termed an element. A vector is a special matrix case, having more than one row, albeit only one column.

measurement model is composed of two regression matrices, two variance-covariance matrices among errors of measurement, and one vector representing the endogenous factor.

The structural model comprises two regression matrices, two variance-covariance matrices (one among the exogenous factors and one among the residual errors associated with the endogenous factors), and three vectors representing the exogenous variables, endogenous variables, and errors associated with the endogenous variables, respectively. An explanation of these matrices is now presented.

The Measurement Model[4]

1. Λ_x is a p by m regression matrix that relates m exogenous factors to each of the p observed variables designed to measure them.
2. Λ_Y is a q by n regression matrix that relates n endogenous factors to each of the q observed measures designed to measure them.
3. Θ_δ is a symmetrical p by p variance-covariance matrix among the errors of measurement for the p exogenous observed variables.
4. Θ_ϵ is a symmetrical q by q variance-covariance matrix among the errors of measurement for the q endogenous observed variables.
5. ν is an n by 1 vector of constant intercept terms.[5]

The Structural Model

1. Γ is an m by n regression matrix that relates the m exogenous factors to the n endogenous factors.
2. B is an n by n regression matrix that relates the n endogenous factors to one another.
3. Φ is an m by m symmetrical variance-covariance matrix among the m exogenous factors.
4. Ψ is an n by n symmetrical variance-covariance matrix among the n residual errors for the n endogenous factors.[6]
5. ξ is an m by 1 vector of exogenous factors.
6. η is an n by 1 vector of endogenous factors.
7. ζ is an n by 1 vector of residuals.

Note: in the general model, LISREL does not permit a priori specification of variances and covariances among the endogenous factors; no variance-covariance matrix is therefore identified here. This does not reflect a limitation in the LISREL program; rather, it is inherent in the model's mathematical logic.

---

[4]By convention, matrices are defined according to their number of rows (r) and columns (c); the number of rows is always specified first. This $r \times c$ description of a matrix is termed the order of the matrix.

[5]We need only concern ourselves with this vector when testing for differences in mean structures (see Chapter 8). Otherwise, ν is assumed to equal zero.

[6]These residual terms are referred to as errors in the equation or as residual errors of prediction; the term "residual" is used to distinguish them from errors of measurement associated with the observed variables.

TABLE 1.1. Summary of Matrix and Greek Notation

Greek letter	Full matrix	Program code	Matrix elements	Type
Measurement Model				
Lambda-X	Λ_X	LX	λ_X	Regression
Lambda-Y	Λ_Y	LY	λ_Y	Regression
Theta delta	Θ_δ	TD	θ_δ	Var/cov
Theta epsilon	Θ_ϵ	TE	θ_ϵ	Var/cov
Nu	—	—	ν	Vector
Structural Model				
Gamma	Γ	GA	γ	Regression
Beta	B	BE	β	Regression
Phi	Φ	PH	ϕ	Var/cov
Psi	Ψ	PS	ψ	Var/cov
Xi (or ksi)	—	—	ξ	Vector
Eta	—	—	η	Vector
Zeta	—	—	ζ	Vector

A summary of these matrices and vectors is presented in Table 1.1 with the program coding for each matrix, since the latter is representative of its Greek name.

3. The LISREL CFA Model

3.1. A Comprehensive Explanation of the CFA Model

Specification of the CFA model involves only a portion of the general LISREL model noted earlier. Furthermore, it is specified either as being exogenous or endogenous; this choice is an arbitrary one.[7] However, once the model is specified as one or the other, all components of the model must be consistent with this specification. In other words, with a CFA model, the researcher works either with an all-X (exogenous) or all-Y (endogenous) model. In hypothesizing a CFA model, the researcher makes specifications with respect to each of the following:

(a) The number of factors (ξs or ηs).
(b) The number of observed variables (Xs or Ys).
(c) Relations between the observed variables and the latent factors (λ_Xs or λ_Ys).

---

[7] This distinction is made here for purposes of simplicity and clarity only. Technically speaking, in CFA models there is no designation of variables as either exogenous or endogenous since there is no specification of causal relations among the latent variables; consequently, ζ, the residual error term associated with the prediction of η from ξ, is zero. This accounts for the freedom on the part of the researcher to elect usage of one or the other model.

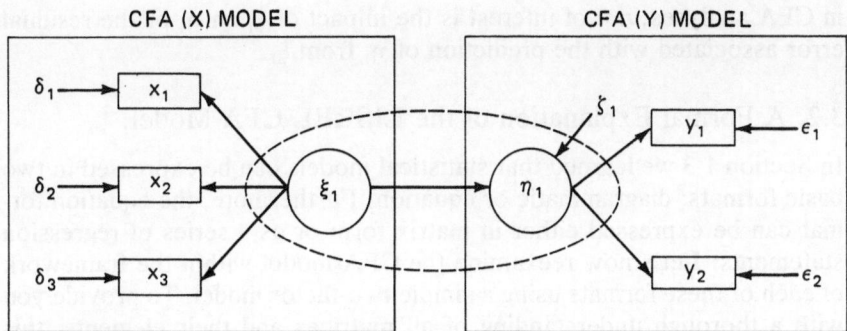

FIGURE 1.1. The LISREL CFA Model Relative to the LISREL Full Model.

(d)Factor variances and covariances (Φ).

(e)Error variances (and possibly covariances) associated with the observed variables (Θ_δ or Θ_ϵ).

To get a visible perspective of the CFA model relative to the general model, let's examine Figure 1.1.

A third requirement in learning to use LISREL is to understand the geometric symbolism depicted in schematic models. The symbols used in Figure 1.1 are defined here.

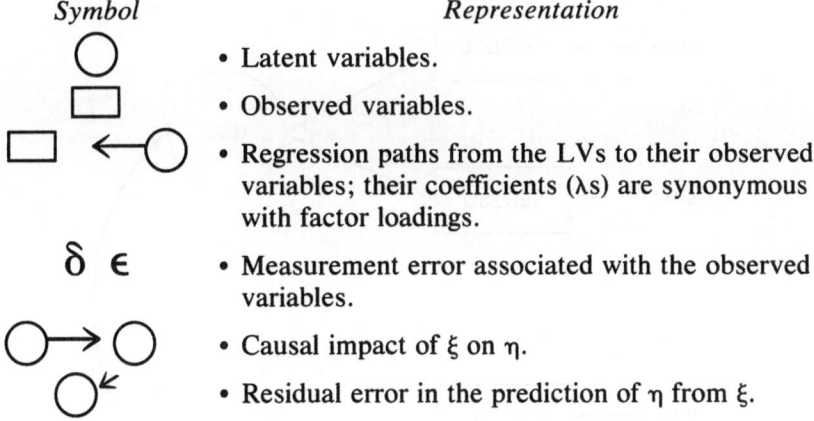

Symbol	*Representation*
○	• Latent variables.
□	• Observed variables.
□ ← ○	• Regression paths from the LVs to their observed variables; their coefficients (λs) are synonymous with factor loadings.
δ ϵ	• Measurement error associated with the observed variables.
○→○	• Causal impact of ξ on η.
○	• Residual error in the prediction of η from ξ.

Within the framework of the model in Figure 1.1, we see two CFA models—one exogenous (X-model) and one endogenous (Y-model). The CFA X-model is a one-factor model measured by three observed variables, while the CFA Y-model is a one-factor model measured by two observed variables. In either case, the factor, its regression on the observed variables, and the errors of measurement are of primary interest in CFA analyses; *not* of interest is the impact of ξ_1 on η_1 or the residual error associated with the prediction of η_1 from ξ_1.

3.2. A Formal Explanation of the LISREL CFA Model

In Section 1.3 we learned that statistical models can be expressed in two basic formats: diagrammatic or equation. Furthermore, the equation format can be expressed either in matrix form or as a series of regression statements. Let's now reexamine the CFA model within the framework of each of these formats using a simple two-factor model. To provide you with a thorough understanding of all matrices and their elements, this model will be expressed first as an all-X model, and then as an all-Y model since examples of both are found in the literature.

Suppose that we have a two-factor model of self-concept: Let the two factors be general self-concept (GSC) and academic self-concept (ASC). Suppose that each factor has two observed variables: Let the two measures of GSC be the General Self subscale of the Self Description Questionnaire (SDQGSC; Marsh & O'Neill, 1984), and the Self-esteem Scale (SESGSC; Rosenberg, 1965). Let the two measures of ASC be the Academic Self-concept subscale of the Self Description Questionnaire (SDQASC) and the Self-concept of Ability Scale (SCAASC); Brookover, 1962).

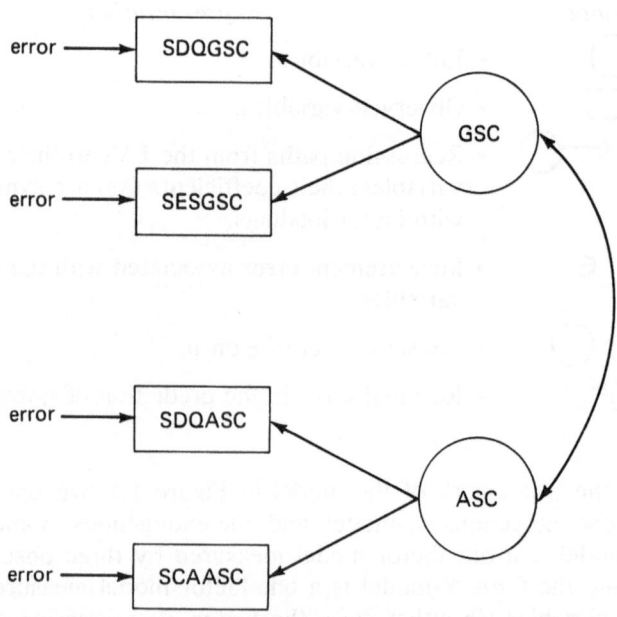

FIGURE 1.2. Hypothesized CFA Model of Self-Concept.

A schematic presentation of this model is shown in Figure 1.2. Here, then, we have a two-factor model consisting of GSC and ASC, with each factor being measured by two observed variables. The observed measures for GSC are SDQGSC and SESGSC; for ASC they are SDQASC and SCAASC. The curved two-headed arrow indicates that GSC and ASC are correlated.

Now, let's translate this model into LISREL notation and reexamine it in terms of a schematic presentation and in terms of a set of equations. We'll look first at the all-X model.

(a) Expressed in schematic form (as in Figure 1.3), ξ_1 and ξ_2 represent GSC and ASC, respectively; the curved arrow indicates that they are correlated. λ_{11} and λ_{21} represent the regression of ξ_1 on X_1 and X_2, respectively; similarly, λ_{32} and λ_{42} represent the regression of ξ_2 on X_3 and X_4, respectively. X_1 and X_2 represent SDQGSC and SESGSC, the observed measures of GSC; X_3 and X_4 represent SDQASC and SCAASC, the observed measures of ASC. δ_1 to δ_4 represent errors of measurement associated with SDQGSC to SCAASC, respectively.

(b) Expressed in equation form:
   (i) As a series of regression equations, Figure 1.4 holds.
   (ii) In matrix form, Figure 1.5 holds.

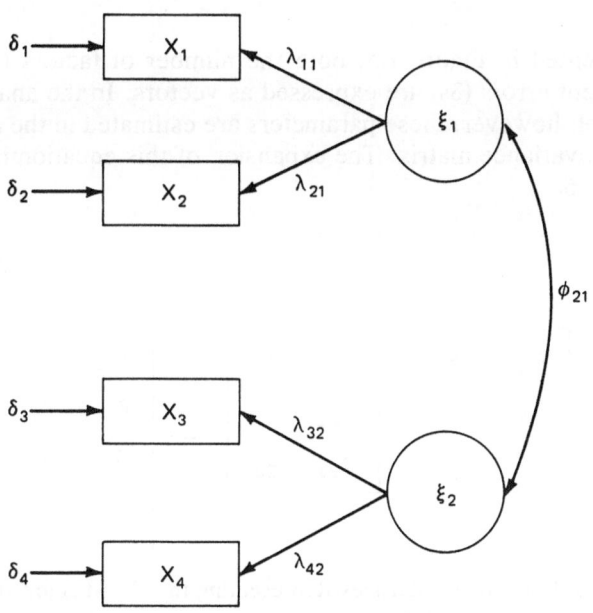

FIGURE 1.3. LISREL All-X CFA Model of Self-Concept.

$$X_1 = \lambda_{11} \xi_1 + \delta_1$$

$$X_2 = \lambda_{21} \xi_1 + \delta_2$$

$$X_3 = \lambda_{32} \xi_2 + \delta_3$$

$$X_4 = \lambda_{42} \xi_2 + \delta_4$$

FIGURE 1.4. Series of Regression Equations Representing the Two-Factor Model of Self-Concept Shown in Figure 1.3.

$$X \quad = \quad \Lambda_X \quad \xi \quad + \quad \delta$$

$$\begin{bmatrix} X_1 \\ X_2 \\ X_3 \\ X_4 \end{bmatrix} = \begin{bmatrix} \lambda_{11} & 0 \\ \lambda_{21} & 0 \\ 0 & \lambda_{32} \\ 0 & \lambda_{42} \end{bmatrix} \begin{bmatrix} \xi_1 \\ \\ \xi_2 \end{bmatrix} + \begin{bmatrix} \delta_1 \\ \delta_2 \\ \delta_3 \\ \delta_4 \end{bmatrix}$$

FIGURE 1.5. Equation of Vectors (Λ always remains in matrix form with the specification of two or more factors) Representing the Two-Factor Model of Self-Concept Shown in Figure 1.3.

As presented in Figure 1.5, both the number of factors (ξs) and the measurement errors (δs) are expressed as vectors. In the analysis of the CFA model, however, these parameters are estimated in the appropriate variance-covariance matrix. The expansion of this equation is presented in Figure 1.6.

$$X \quad = \quad \Lambda_X \quad \Phi \quad + \quad \Theta_\delta$$

$$\begin{bmatrix} X_1 \\ X_2 \\ X_3 \\ X_4 \end{bmatrix} = \begin{bmatrix} \lambda_{11} & 0 \\ \lambda_{21} & 0 \\ 0 & \lambda_{32} \\ 0 & \lambda_{42} \end{bmatrix} \begin{bmatrix} \phi_{11} & \\ \phi_{21} & \phi_{22} \end{bmatrix} + \begin{bmatrix} \theta_{11} & & & \\ 0 & \theta_{22} & & \\ 0 & 0 & \theta_{33} & \\ 0 & 0 & 0 & \theta_{44} \end{bmatrix}$$

FIGURE 1.6. Equation of Matrices Representing the Two-Factor Model of Self-Concept Shown in Figure 1.3.

3.3. Additional Points of Explanation

1. By convention, the numbering of matrix elements is such that the first number represents the row, while the second represents the column. Thus ϕ_{21} indicates an element in the second row, first column of the matrix Φ.

2. The Λ matrix is often referred to as the factor-loading matrix because it demonstrates the pattern of how each observed variable is linked to each factor. For example, λ_{11} and λ_{21} indicate that the first two elements in the regression matrix represent X_1 and X_2, respectively, and that both load on Factor 1 (GSC); the two zeros in the same column indicate that no other variables load on Factor 1.[8] Of course the reverse pattern holds for Factor 2.

3. Recall from Section 2.2 that the factor variance-covariance matrix for the ξs was identified as the Φ matrix. Thus, while the vector of ξs indicated that there were two factors, the variance for these factors (ϕ_{11}, ϕ_{22}) and the covariance between them (ϕ_{21}) represent elements in the Φ matrix.[9]

4. Recall from Section 2.2 that the error variance-covariance matrix for the δs was the Θ_δ matrix. Thus, the error variances are represented by θ_{δ_1} to θ_{δ_4}, respectively. One assumption underlying the CFA model is that errors of measurement are assumed to be uncorrelated; the zeros therefore indicate that no parameters representing error covariances will be estimated. More details regarding assumptions will be addressed in Chapter 2.

5. Recall from Section 2.2 that the variance-covariance matrix for the residual terms (ζ) was identified as Ψ in the general model. However, since the CFA model does not include causal relations between ξ and η, the two latent factors, the residual is reduced to zero. Thus, correlations among the latent factors in the all-Y model (ηs) are estimated in the Ψ matrix.

Let's now examine the same two-factor model expressed as an all-Y model

(a) Expressed in schematic form, it would look like Figure 1.7.

---

[8] Note that if this had been a one-factor model, the Λ and Φ matrices would have been expressed as vectors.

[9] Had no correlation been specified between the two factors (i.e., they were considered to be orthogonal), the ϕ_{21} would have been specified as zero—

$$\begin{bmatrix} \phi_{11} & \\ 0 & \phi_{22} \end{bmatrix}.$$

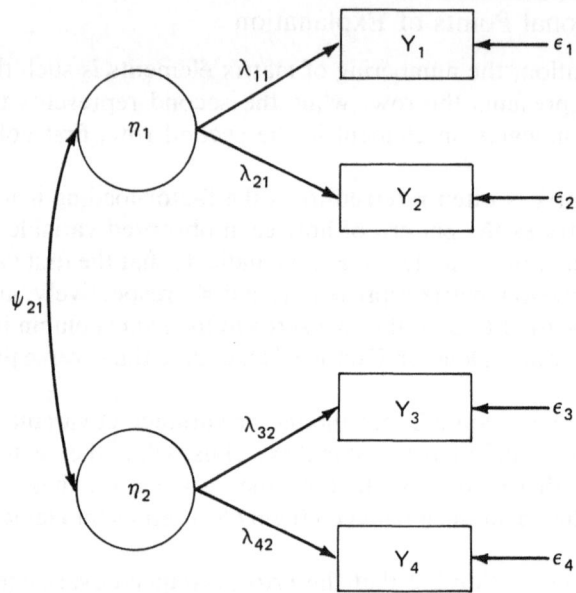

FIGURE 1.7. LISREL All-Y CFA Model of Self-Concept.

(b)Expressed in equation form, it would look like one of the following:
   (i) As a series of regression equations (Figure 1.8).
   (ii) In matrix form (ηs and ϵs as vectors, Figure 1.9).
   (iii) In expanded matrix form (Figure 1.10).

$$Y_1 = \lambda_{11} \eta_1 + \epsilon_1$$

$$Y_2 = \lambda_{21} \eta_1 + \epsilon_2$$

$$Y_3 = \lambda_{32} \eta_2 + \epsilon_3$$

$$Y_4 = \lambda_{42} \eta_2 + \epsilon_4$$

FIGURE 1.8. Series of Regression Equations Representing the Two-Factor Model of Self-Concept Shown in Figure 1.7.

$$
Y \quad = \quad \Lambda_Y \quad \eta \quad + \quad \varepsilon
$$

$$
\begin{bmatrix} Y_1 \\ Y_2 \\ Y_3 \\ Y_4 \end{bmatrix}
=
\begin{bmatrix} \lambda_{11} & 0 \\ \lambda_{21} & 0 \\ 0 & \lambda_{32} \\ 0 & \lambda_{42} \end{bmatrix}
\begin{bmatrix} \eta_1 \\ \eta_2 \end{bmatrix}
+
\begin{bmatrix} \varepsilon_1 \\ \varepsilon_2 \\ \varepsilon_3 \\ \varepsilon_4 \end{bmatrix}
$$

FIGURE 1.9. Equation of Vectors (except for Λ_y) Representing the Two-Factor Model of Self-Concept Shown in Figure 1.7.

$$
Y \quad = \quad \Lambda_Y \quad \Psi \quad + \quad \Theta_\epsilon
$$

$$
\begin{bmatrix} Y_1 \\ Y_2 \\ Y_3 \\ Y_4 \end{bmatrix}
=
\begin{bmatrix} \lambda_{11} & 0 \\ \lambda_{21} & 0 \\ 0 & \lambda_{32} \\ 0 & \lambda_{42} \end{bmatrix}
\begin{bmatrix} \psi_{11} & \\ \psi_{21} & \psi_{22} \end{bmatrix}
+
\begin{bmatrix} \theta_{11} & & & \\ 0 & \theta_{22} & & \\ 0 & 0 & \theta_{33} & \\ 0 & 0 & 0 & \theta_{44} \end{bmatrix}
$$

FIGURE 1.10. Equation of Matrices Representing the Two-Factor Model of Self-Concept Shown in Figure 1.7.

4. Summary

This chapter provided a basic working knowledge of LISREL symbols and modeling procedures. An explanation of basic concepts related to LISREL modeling was presented first; these included the distinction between observed and latent variables, the underlying purposes of statistical modeling in general, and factor analytic modeling in particular. The composition of the general LISREL model was presented next, with the components of the model being presented both mathematically and schematically; related LISREL notation was introduced. Finally, a detailed explanation of the LISREL CFA model was demonstrated with both all-X and all-Y model specifications.

2
Using the LISREL Program

The purpose of this chapter is to introduce you to the general format of the LISREL program. Expanded comments regarding various aspects of the input and output will be addressed in subsequent chapters that focus on specific CFA model applications. Since this book is intended only as a guide to using LISREL, it is limited to applications based on maximum likelihood estimation. For applications based on other models and/or use of other estimation techniques, the reader should refer to the LISREL manual. In general, however, the reader is encouraged to use this book in conjunction with the LISREL manual to ensure a thorough understanding of the link between CFA modeling and the LISREL program.

The basic elements of the LISREL program are now outlined. We will first examine specifications regarding the input of information and then we will review the information provided on a standard output.

LISREL Input

1. Basic Rules

1.1. Keywords

The LISREL program is controlled by two-letter keywords that represent both card and parameter names. Although these names may contain any number of letters, only the first two are recognized by LISREL; all keywords are separated by spaces or commas.

1.2. Control Cards

In order for the LISREL program to run, it must be provided with data, along with four important pieces of information: title of problem run, data specification (i.e., description of data being entered), model specification (i.e., description of model to by analyzed), and output specification (i.e., type of information desired). This information is provided through four basic control cards that must be included for every problem run; these are as follows:

1. TItle card.
2. DAta card.
3. MOdel card.
4. OUtput card.

The capitalized letters for each card represent the required keywords. Except for the TI card, additional information is required within each of the other categories in order to complete the specification requirements. This information is provided by parameter specifications and additional control cards. These four control cards require a fixed input position, as listed earlier.

Now let's examine, in more detail, each of these four major control cards, their parameters, and supplementary control cards. For simplicity, examples of input setups will be based on the model presented in Chapter 1 (see Figure 1.2); the sample size is arbitrarily chosen to be 400.

2. Problem Run Specification

The first card for each problem run must specify a title for the problem; this is specified on the TI control card. Although there can be more than one card (i.e., more than one line) of input, most users find it convenient to limit the title to less than 79 characters (i.e., 79 columns); column 80 must be left blank. If more than one card is used, however, only column 80 on the last card needs to be left blank; the first 79 characters of the title will be printed on each page of the output.

> e.g. TI Multidimensionality of Self-concept

3. Data Specification

The second major control card is the DA card, which defines the data to be analyzed. In order to provide these details, however, additional information is provided by means of the following specifications:

NGroups = Number of groups for which data are available.

NInpvar = Number of input variables.

NObs   = Number of observations (i.e., sample size); if N is unknown and raw data are being input, set NO = 0; the program will compute N.

MAtrix = Type of matrix to be analyzed.
- KM for correlation matrix.
- CM for covariance matrix.
- MM for moment matrix.
- AM for augmented moment matrix.

It is important to note, however, that the matrix to be analyzed may differ from the one read into the computer as data input. For example, the data may be input in the form of a KM, but you may wish to have analyses based on the CM. LISREL can always compute the matrix to be analyzed, regardless of the type of data that are input.

> e.g. DA NG = 1 NI = 4 NO = 400 MA = KM
>   ↓
> Cols 1 2 4 5 25

In this example, the DA card indicates that there is only one group to be analyzed,[1] there are four input variables, the sample size is 400, and the data are going to be entered in the form of a correlation matrix.

3.1. Further Specification of the Data

The DA card is further defined by four additional pieces of information; each is entered on a separate line of input as follows:

(i) LAbels
    ↑
    Col 1
    One label, eight characters or less, must be provided for each input variable. The easiest way to provide this information is to enter the variable names in free format. As such, they are entered in the same order as the variables appear in the data set; each label is enclosed in single quotes and separated by blanks. This format is indicated in column 1 of the next card.
(ii) Format in which variable labels are to be read. If the format is free, as noted earlier, an asterisk (*) is placed in column 1.
(iii) The matrix form in which the data are being input. Two pieces of information are required: the type of matrix and the particular form of the matrix. Each piece of information is represented by a two-letter keyword; both are entered on the same line, separated by a blank. The choices are as follows:

Matrix Type	Matrix Form
RA for raw data matrix.	FU for full matrix.
KM for correlation matrix.	SY for symmetric matrix.
CM for covariance matrix.	
MM for moment matrix.	

---

[1]The default value for NG is 1.0. This means that it is not necessary to enter NG = 1; LISREL will automatically base the analyses on one group. The input would then read: DA NI = 4 NO = 400 MA = KM.

(iv) The information in the input format tells LISREL how each set of numbers represents the variables in the data set. The input format can be either fixed or free. In either case, LISREL reads all data row-wise, from left to right.

Fixed Format. This format indicates that the data are to be read according to a specific formula, a Fortran statement. It specifies the number and location of columns occupied by each variable, in addition to the number of decimal points included for each variable score, if any. A Fortran statement is always parenthesized; the initial parenthesis is placed in column 1.

Free Format. This format indicates that the data are to be read as one long string of numbers; no specific columns are linked to any variable. The only requirement here is that each variable score must be separated by a blank. When this type of format is used, an asterisk (*) is placed in Column 1.

The above four lines are then followed by data entry.

Continuing with our example based on Figure 1.2, let's look at the input of this additional information using a Fortran statement and a free format.

```
e.g. (a) LA
         *
         'SDQGSC' 'SESGSC' 'SDQASC' 'SCAASC'
         KM SY
         (4F3.2)
         100
          70100
          50 55100
          45 48 80100
```

This format indicates that there are four rows representing four variables in a symmetric matrix form. LISREL then reads each row from left to right, counting three digits and then placing a decimal point to the left of the last two digits. Blanks represent zeros and are therefore taken into consideration in counting the three digits.

```
e.g. (b) LA
         *
         'SDQGSC' 'SESGSC' 'SDQASC' 'SCAASC'
         KM SY
         *
         1.00
          .70 1.00
          .50 .55 1.00
          .45 .48 .80 1.00
```

Here we have the same matrix of numbers but read in free format fashion. In this form, blanks are used to separate score values.

• *Hint*
Remember to account for column 1 when formulating the Fortran statement. Otherwise, you will get an error message that will in no way alert you to the fact that this has not been done (e.g., the input matrix is not positive definite). When using the free format, be sure to enter the first 1.00 in column 1.

3.2. Optional Input

Means and Standard Deviations. For certain problems, it may be necessary to include the mean and standard deviation values for each variable. This information can be included by first adding an ME or SD card, followed by a free format card (*), followed by the actual values for each variable. In other words, a minimum of three lines is required for the input of means and for the input of standard deviations. This information is entered, beginning with column 1. When both means and standard deviations are added, the three lines related to means are added first, as shown later. This information follows immediately after entry of the data matrix.

```
e.g. ME
     *
     76.41 52.90 55.60 49.22
     SD
     *
     10.10 9.05 4.56 7.80
```

Selection of Variables. For a variety of reasons, the user may wish to use only certain variables from those listed on the LA card. This is easily done by adding a SElect card after the row of SDs and listing the variables (either by name or number) in the order in which they are to be read; this is followed by a slash (/); the slash indicates that certain variables are being eliminated from the analyses.

Suppose that in our previous example we did not wish to include the variable 'SESGSC' in the analyses. We then eliminate it by entering a select card as:

SE 1 3 4/

The SE card is also used to change the order in which the variables are read into the analyses. In the previous example, the order of the variables is being read automatically in the order, 1 2 3 4. Therefore, an SE card is not required. But if the cards are to read in a different order than the way they have been input, an SE card is required. For example,

SE 1 3 2 4

This indicates that the variable 'SDQASC' is to be read after 'SDQ-GSC' instead of after 'SESGSC.'

• *Hint*
>   1. If all variables are being used, the SE card is not required.
>   2. If the SE card has been used to indicate missing variables, and the slash has not been included, LISREL will print an error message. However, the error message may not relate in any way to the fact that the slash has been omitted. The fact that the problem did not run should alert you to the fact that something is not right.

4. Model Specification

The third major control card is the MO card, which specifies the model to be analyzed. Model specification involves four pieces of information:

1. The number of observed variables in the model (Xs or Ys).
2. The number of latent variables in the model (ξs or ηs). LISREL reports ξs as KSIs, and the ηs as ETAs.
3. The form of each matrix to be analyzed.
4. The estimation status of each matrix.

We turn now to each of these components.

4.1. Observed Variables

Since we are only interested in CFA models as presented in Chapter 1, model specifications will include either X variables or Y variables—but not both. The keywords are as follows:

> NX = number of X variables in the model.
> NY = number of Y variables in the model.

4.2. Latent Variables

As noted in Chapter 1, a CFA model specification will include either ξs (KSIs) or ηs (ETAs)—not both. The keywords are as follows:

> NK = number of ξs in the model.
> NE = number of ηs in the model.

4.3. Basic Matrix Forms

To understand the CFA applications presented in this book, five basic matrix forms need to be known; these are presented in Figure 2.1.

$$
\begin{bmatrix}
X & X & X & X \\
X & X & X & X \\
X & X & X & X \\
X & X & X & X
\end{bmatrix}
$$

Full Matrix

$$
\begin{bmatrix}
X & & & \\
X & X & & \\
X & X & X & \\
X & X & X & X
\end{bmatrix}
\qquad
\begin{bmatrix}
X & & & \\
& X & & \\
& & X & \\
& & & X
\end{bmatrix}
$$

Symmetric Matrix   Diagonal Matrix

$$
\begin{bmatrix}
1 & 0 & 0 & 0 \\
0 & 1 & 0 & 0 \\
0 & 0 & 1 & 0 \\
0 & 0 & 0 & 1
\end{bmatrix}
\qquad
\begin{bmatrix}
0 & 0 & 0 & 0 \\
0 & 0 & 0 & 0 \\
0 & 0 & 0 & 0 \\
0 & 0 & 0 & 0
\end{bmatrix}
$$

Identity Matrix   Zero Matrix

FIGURE 2.1. Basic Matrix Forms in LISREL.

Several points need to be noted here with respect to the treatment of these matrices in the LISREL program. These are as follows:

1. The Xs in the FU, SY, and DI matrices represent values that are stored in the computer. This means that:
   (a) If a matrix is specified as SY, one cannot refer to elements in the upper triangle of the matrix. As far as LISREL is concerned, these elements do not exist.
   (b) If a matrix is specified as DI, one cannot refer to off-diagonal elements. Again, these elements are not stored in the computer.
2. If a matrix is stored as an ID or ZE matrix, one cannot refer to any element in these matrices. This is because neither matrix is stored in the computer.

4.4. Matrix Estimation Status

As shown in Figure 1.3, if the model to be analyzed is an all-X CFA model, then only the lambda-X (LX), phi (PH), and theta-delta (TD) matrices are of interest. Alternatively, if an all-Y model is specified, then the lambda-Y (LY), psi (PS), and theta-epsilon (TE) matrices are of interest. In either case, the status of each of the three matrices must be specified. This means that LISREL parameters representing these matrices must be specified according to whether or not they are to be estimated, and if they are, then how they are to be estimated. That is, they are specified as being FRee, FIxed, or EQual to other parameters in the model.

FRee parameters indicate that these values are unknown and therefore will be estimated by the program.

FIxed parameters are assigned some particular value by the investigator; they are therefore not estimated by the program.

EQual parameters indicate that they have been constrained to have the same estimated value as certain other parameters in the model. LISREL estimates the initial parameter; all other parameters that are constrained equal to it will thus have the same estimated value.

Specification of the status of a matrix means that all elements in that matrix have the same status.

> e.g. LX = FU,FI specifies that Lambda-X is a full matrix with all elements fixed to some value, to be input later in the setup. The default value for fixed parameters is 0.0. This means that if no value is specified by the investigator, the value will be automatically fixed at 0.0 by the program.

> e.g. TD = DI,FR specifies that Theta Delta is a diagonal matrix with all elements free to be estimated by the program.

Before proceeding, we need to make a slight digression in order to say a few words about Default Values. All computer programs operate with certain default values. What this means is that when a particular specification is omitted from an input setup, the program automatically implements a fixed value that has been preselected on the basis of most common use. With LISREL, default values are associated with particular matrix specifications.

The advantage of default values is that they can often reduce, substantially, the amount of time required to input the specification information. In order not to confuse you at this point, however, example inputs will specify all parameters and will not rely on the default values.

Default values for the specification of parameter matrices for CFA models are presented in Table 2.1. Thus, it can be seen that if LX only appeared on the MO card, the program would, by default, treat it as a full matrix with elements fixed to 0.0 (unless otherwise specified later in the input).

TABLE 2.1. Default Values for Parameter Matrices in LISREL CFA Models

Matrix name	Greek notation	LISREL notation	Default form	Default mode (fixed/free)
Lambda-X	Λ_x	LX	FU	FI
Lambda Y	Λ_y	LY	FU	FI
Phi	Φ	PH	SY	FR
Psi	Ψ	PS	SY	FR
Theta-delta	Θ_δ	TD	DI	FR
Theta-epsilon	Θ_ϵ	TE	DI	FR

4.5. Additional Model Specification

In Section 4.4., we learned how to specify the status of an entire matrix. However, in most cases, the model will require further refinement. This is accomplished by specifying certain matrix elements as fixed, free, or constrained equal to other parameters. As such, three additional control cards come into play. We turn now to these input details.

The Estimation Status of Matrix Elements. Regardless of the status of a matrix, any of its elements can be specified to have a different status. That is to say, although the matrix may be specified as fixed (free), any of its elements may be individually specified as free (fixed). Matrix elements are identified by parenthesizing their coordinate points (i.e., their intersecting row and column numbers as shown in Chapter 1), which are separated by commas.

> e.g. MO NX = 4 NK = 2 LX = FU,FI PH = SY,FR TD = DI,FR
>   FR LX(2,1) LX(4,2)
>   FI TD(1,1)

This example specifies that the model has four X (observed) variables and two ξ (latent) variables. LX is a full matrix with its elements fixed to some value; two elements, however, are to be freely estimated. This specification is depicted in Figure 2.2

$$
\begin{bmatrix}
\text{MO} \quad \text{NX} = 4 \quad \text{NK} = 2 \quad \text{LX} = \text{FU,FI} \\
\text{FR} \quad \text{LX(2,1)} \quad \text{LX(4,2)}
\end{bmatrix}
$$

$$
\text{LX}
\begin{array}{cc}
\xi_1 & \xi_2 \\
\begin{bmatrix}
\text{LX}_{11} \ (\text{FI}) & \text{LX}_{12} \ (\text{FI}) \\
\text{LX}_{21} \ (\text{FR}) & \text{LX}_{22} \ (\text{FI}) \\
\text{LX}_{31} \ (\text{FI}) & \text{LX}_{32} \ (\text{FI}) \\
\text{LX}_{41} \ (\text{FI}) & \text{LX}_{42} \ (\text{FR})
\end{bmatrix}
\end{array}
\qquad
\Lambda_x
\begin{array}{cc}
\xi_1 & \xi_2 \\
\begin{bmatrix}
\lambda_{11} \ (\text{fixed}) & \lambda_{12} \ (\text{fixed}) \\
\lambda_{21} \ (\text{free}) & \lambda_{22} \ (\text{fixed}) \\
\lambda_{31} \ (\text{fixed}) & \lambda_{32} \ (\text{fixed}) \\
\lambda_{41} \ (\text{fixed}) & \lambda_{42} \ (\text{free})
\end{bmatrix}
\end{array}
$$

(a) Matrix with LISREL Notation   (b) Matrix with Greek Notation

$$
\Lambda_x \quad
\begin{array}{cc}
\xi_1 & \xi_2 \\
\begin{bmatrix}
1.0 & 0.0 \\
\lambda_{21} & 0.0 \\
0.0 & 1.0 \\
0.0 & \lambda_{42}
\end{bmatrix}
\end{array}
$$

(c) Matrix with Assigned Values for Fixed Parameters

FIGURE 2.2. The Specified LX Matrix.

The assigned values in Figure 2.2(c) need some elaboration. Since we are working with a CFA model, we are postulating that certain observed variables will load on particular latent variables. In this case, we are hypothesizing that λ_{11} and λ_{21} load on Factor 1 (ξ_1), and that λ_{32} and λ_{42} load on Factor 2 (ξ_2). Thus, λ_{31}, λ_{41}, λ_{12}, and λ_{22} are fixed to zero. For purposes of statistical identification[2] and in order to establish the scale of metric, one of the free parameters being estimated for each factor should be fixed to 1.00. Although most investigators fix the first of a set of λs to 1.00, this decision is an arbitrary one.

Similarly, TD is specified as a diagonal matrix with all elements except TD(1,1) free to be estimated. As shown in Figure 2.3, only the diagonal elements are of interest, therefore the off-diagonal parameters have been fixed to zero; δ_{11} has been fixed to .20.

$$
\begin{bmatrix}
\text{MO} & \text{NX} = 4 & \text{NK} = 2 & \text{TD} = \text{DI,FR} \\
\text{FI} & \text{TD(1,1)} & &
\end{bmatrix}
$$

$$
\text{TD} \quad
\begin{array}{cccc}
X_1 & X_2 & X_3 & X_4 \\
\begin{bmatrix}
\text{TD}_{11}\ (\text{FI}) & \text{TD}_{12}\ (\text{FI}) & \text{TD}_{13}\ (\text{FI}) & \text{TD}_{14}\ (\text{FI}) \\
\text{TD}_{21}\ (\text{FI}) & \text{TD}_{22}\ (\text{FR}) & \text{TD}_{23}\ (\text{FI}) & \text{TD}_{24}\ (\text{FI}) \\
\text{TD}_{31}\ (\text{FI}) & \text{TD}_{32}\ (\text{FI}) & \text{TD}_{33}\ (\text{FR}) & \text{TD}_{34}\ (\text{FI}) \\
\text{TD}_{41}\ (\text{FI}) & \text{TD}_{42}\ (\text{FI}) & \text{TD}_{43}\ (\text{FI}) & \text{TD}_{44}\ (\text{FR})
\end{bmatrix}
\end{array}
$$

(a) Matrix with LISREL Notation

FIGURE 2.3. The Specified TD Matrix.

---

[2]A discussion of identification is beyond the scope of this book; for an extensive discussion of this topic see e.g., Long, 1983; Saris & Stronkhorst, 1984.

	X_1	X_2	X_3	X_4
Θ_δ	δ_{11} (fixed)	δ_{12} (fixed)	δ_{13} (fixed)	δ_{14} (fixed)
	δ_{21} (fixed)	δ_{22} (free)	δ_{23} (fixed)	δ_{24} (fixed)
	δ_{31} (fixed)	δ_{32} (fixed)	δ_{33} (free)	δ_{34} (fixed)
	δ_{41} (fixed)	δ_{42} (fixed)	δ_{43} (fixed)	δ_{44} (free)

(b) Matrix with Greek Notation

	X_1	X_2	X_3	X_4
Θ_δ	0.2	0.0	0.0	0.0
	0.0	δ_{22}	0.0	0.0
	0.0	0.0	δ_{33}	0.0
	0.0	0.0	0.0	δ_{44}

(c) Matrix with Assigned Values for Fixed Parameters

FIGURE 2.3. Continued.

Alternatively, the preceding example could be specified as follows:

```
e.g. MO NX = 4 NK = 2 LX = FU,FR PH = SY,FR
     TD = DI,FI
     FI LX(1,1) LX(3,1) LX(4,1) LX(1,2) LX(2,2) LX(3,2)
     FR TD(2,2) TD(3,3) TD(4,4)³
```

A review of Figures 2.2 and 2.3 will quickly demonstrate that the two model specifications are identical.

• *Hint*
It is now easy to see that there is more than one way to specify a model. For the sake of expedience, it is best to specify the matrix in accordance with the desired status of most of its elements. In other words, if most of the elements are to be estimated, specify the status of the matrix as 'free.' On the other hand, if most elements are fixed parameters (see, e.g., Figures 2.2 and 2.3), specify the matrix as 'fixed.' In this way, less input is required in specifying the status of individual matrix elements.

---

[3]Where a consecutive range of values is to be estimated, the first and last elements can be hyphenated, e.g., TD(2,2)-TD(4,4).

Assigning Values for Fixed Parameters. In the case where a fixed parameter is to have a nonzero value rather than the default value of 0.0, the
VAlue and/or STart control cards are used. The VA and ST cards are
equivalent, and may be used synonymously. The keyword on either card
is followed by a number, which is taken as the assigned value for the
accompanying list of matrix elements.

```
e.g. FI  LX(1,1) LX(3,2)
     VA  1.0 LX(1,1) LX(3,2)
or   ST  1.0 LX(1,1) LX(3,2)
```

This example indicates that elements (1,1) and (3,2) in the LX matrix
are to be fixed to a value of 1.0.

• *Hint*
   For convenience, it is often advisable to use the VA cards to indicate
   assigned values for fixed parameters, and ST cards to indicate starting
   values for free parameters.

Beginning with the LISREL V version, the program can generate its
own start values. However, the user has the option to enter values if he
or she wishes. In this case, the ST control card is used to specify starting
values for the estimation of free rather than fixed parameters.

However, readers are urged not to rely too heavily on the program-
generated start values. This feature appears to work well with simple
models that have estimated values close to the initial LISREL estimates.
However, as soon as model specifications become more complex or when
estimates are not close to the initial LISREL estimates, the program often
abends (i.e., terminates prematurely).

It is recommended, therefore, that the user always enters his or her
own start values. These values may be derived from a preliminary run
in which only the initial estimates are requested. However, a quick and
reasonable rule of thumb for a set of CFA start values is as follows: λs
= 7.00; ϕs or ψs (variances) = .50; ϕs or ψs (covariances) = .20; δs or
ϵs (variances) = .10.

• *Hints*
   In selecting start values:
   1. Be sure to make diagonal values larger than off-diagonal values or
   you will get an error message that "the information matrix is not positive definite."
   2. If the program abends, consider the possibility of negative start values for some of the λs and ϕs (covariances).

5. Output Specification

Decisions regarding output fall into two categories: method of parameter estimation and information related to analyses. This information is specified on the OU card.

5.1. Method of Parameter Estimation

The estimation of parameters may be obtained by five different methods; these options involve the use of: instrumental variables (IV), two-stage least squares (TSLS), unweighted least squares (ULS), generalized least squares (GLS), and maximum likelihood (ML). Since the underlying assumptions differ for each of these methods, the user is strongly advised to select the one most appropriate for his or her data. The IV and TSLS methods are fast and are not based on an iterative process; they can be used conveniently with large samples. The ULS, GLS, and ML methods, on the other hand, compute estimates iteratively, using the IV and TSLS estimates as start values; these constitute the LISREL automatic start values.

The user selects any one of the five estimation procedures and enters the keyword on the OU card. In the selection of ULS, GLS, and ML, LISREL prints the initial and the final estimates. The keywords and the resulting output are as follows:

IV = only the IV estimates are computed.
TS = only the TSLS estimates are computed.
UL = both the IV and ULS estimates are computed.
GL = both the TSLS and GLS estimates are computed.
ML = both the TSLS and ML estimates are computed.

> *Default* = ML
> This means that if no estimation method is entered on the OU card, parameters will automatically be estimated using ML. Initial estimates based on IV are also provided.

• *Hint*
  If the user wants to enter his or her own start values but has no idea of values, it can be helpful to obtain these from the program by requesting only the initial estimates. This can be done by entering the keyword IV on the OU card; this is equivalent to TS. If more than one of these keywords is entered on the OU card, LISREL will only recognize the last one.

5.2. Information Related to Analyses

Although LISREL provides a standard output (see Figure 2.4), the user can select from a number of options regarding additional information to be printed on the output. The appropriate keyword is simply added to the OU card. The options available, together with the accompanying keywords, are as follows:

PT = Print technical output.
SE = Print standard errors.
TV = Print *t*-values.
PC = Print correlations of estimates.
RS = Print $\hat{\Sigma}$, residuals S-$\hat{\Sigma}$, normalized residuals, Q-plot.
EF = Print total effects.
VA = Print variances and covariances.
MR = Equivalent to RS, EF, and VA.
MI = Print modification indices.
FS = Print factor scores regression.
FD = Print first derivatives.
SS = Print standardized solution.
AL = Print all output.
NS = No automatic start values.
TO = Print with 80 characters per record; default: 132 characters.
ND = Number of decimal places to be printed (0–8); default: ND = 3.
TM = Maximum number of CPU-seconds allowed for problem; default: TM = 60.

The meaning and interpretation of the optional output are discussed as they relate to specific applications as presented in subsequent chapters.

• *Hint*
1. Be sure to add NS to the OU card when entering your own start values or you will get a strange error message that is in no way related to the problem.
2. The LISREL program is expensive in that it uses a lot of CPU time. To cut costs, only request output that is directly of use to you at any one step in your analysis. For example, it only makes sense to request the standardized solution after you have achieved the final best-fitting model.

LISREL Output

1. Standard Output

LISREL provides a standard output that is printed whether or not other selected options have been entered. It includes the following information: log of read control cards, the title, the parameter listing, the parameter specifications, the matrix to be analyzed, the initial estimates, the LISREL estimates (ML or ULS), and the overall goodness-of-fit measures. Figure 2.4 lists the standard output based on our hypothetical two-factor model presented in Figure 1.3.

FIGURE 2.4. LISREL Standard Output for Model in Figure 1.3.

LISREL VI

(1) Confirmatory Factor Analysis

Log read for LISREL Control cards:

DA NI = 4 NK = 2   LX = FU,FI   PH = SY,FR   TD = DI,FI

FR LX(2,1) LX(4,2)

VA 1.0 LX(1,1) LX(3,2)

ST .7 LX(2,1) LX(4,2)

ST .5 PH(1,1) PH(2,2) PH(3,3) PH(4,4)

ST .2 PH(2,1) PH(3,1) PH(4,1) PH(3,2) PH(4,2) PH(4,3)

ST .3 TD(1,1) - TD(4,4)

OU NS

(2) Confirmatory Factor Analysis

Number of input variables	4
Number of Y - variables	0
Number of X - variables	4
Number of ETA - variables	0
Number of KSI - variables	2
Number of observations	400

Model specification

Lambda x   Full, fixed PHI Symm, Free

Theta Delta Diag., Free

Output requested

Technical output	No
Standard Errors	No
T-values	No
Correlations of Estimates	No
Residuals	No

Total Effects	No
Variances and Covariances	No
Modification Indices	No
Factor Scores Regressions	No
First Order Derivatives	No
First Order Derivatives	No
Standardized Solution	No

(3) Confirmatory Factor Analysis

Correlation Matrix to Be Analyzed

	SDQGSC	SESGSC	SDQASC	SCAASC
SDQGSC	1.000			
SESGSC	.701	1.000		
SDQASC	.504	.551	1.000	
SCAASC	.452	.480	.804	1.000

DETERMINANT = 0.520102D-03

(4) Confirmatory Factor Analysis

Parameter Specifications

Lambda X

	KSI 1	KSI 2
SDQGSC	0	0
SESGSC	1	0
SDQASC	0	0
SCAASC	0	2

PHI

	KSI 1	KSI 2
KSI 1	3	
KSI 2	4	5

THETA DELTA

SDQGSC	SESGSC	SDQASC	SCAASC
6	7	8	9

(5) Confirmatory Factor Analysis

Starting Values

Lambda X

	KSI 1	KSI 2
SDQGSC	1.000	0.0
SESGSC	.700	0.0
SDQASC	0.0	1.000
SCAASC	0.0	.700

PHI

	KSI 1	KSI 2
KSI 1	.500	
KSI 2	.200	.500

THETA DELTA

SDQGSC	SESGSC	SDQASC	SCAASC
.100	.100	.100	.100

SQUARED MULTIPLE CORRELATIONS FOR X-VARIABLES

SDQGSC	SESGSC	SDQASC	SCAASC
.700	.700	.700	.700

TOTAL COEFFICIENT OF DETERMINATION FOR X-VARIABLES IS .997.

(6) Confirmatory Factor Analysis

LISREL ESTIMATES (MAXIMUM LIKELIHOOD)

LAMBDA X

	KSI 1	KSI 2
SDQGSC	1.000	0.0
SESGSC	.819	0.0
SDQASC	0.0	1.000
SCAASC	0.0	.973

PHI

	KSI 1	KSI 2
KSI 1	.762	
KSI 2	.209	.657

THETA DELTA

SDQGSC	SESGSC	SDQASC	SCAASC
.238	.192	.306	.119

SQUARED MULTIPLE CORRELATIONS FOR X-VARIABLES

SDQGSC	SESGSC	SDQASC	SCAASC
.762	.511	.808	.694

TOTAL COEFFICIENT OF DETERMINATION FOR X-VARIABLES IS .998.

MEASURES OF GOODNESS OF FIT FOR THE WHOLE MODEL:

CHI-SQUARE WITH 6 DEGREES OF FREEDOM IS 22.57 (PROB. LEVEL = 0.0).

GOODNESS OF FIT INDEX IS 0.875.

ADJUSTED GOODNESS OF FIT INDEX IS 0.701

ROOT MEAN SQUARE RESIDUAL IS 0.032.

2. Error Messages

Errors are inevitable, regardless of how familiar someone is with a computer package. Usually, an error message provides some clue as to the location of the error and how it might be corrected. Unfortunately, this is not always the case with LISREL; as such, the link between the message and the problem is often remote. In most cases, these seemingly bizarre messages are related to simple syntax errors; that is, errors that result from such things as omitting a symbol (e.g., slash on the SE card), omitting a space or comma (e.g., LX = FUFI), or misspelling a keyword.

Thus, the first thing to do when confronted with an error message is to reexamine your input cards, looking very carefully for syntax errors. Some common mistakes you might make are:

- Using keywords that don't conform to the LISREL naming conventions.
- Omitting required slashes, equal signs, commas, or spaces.
- Leaving pairs of parentheses or apostrophes unmatched.
- Placing the TI, DA, MO, or OU cards in the wrong order.
- Inputting a data correlation matrix and forgetting to account for column 1.
- Using lower-case, rather than upper-case lettering.
- Using start values that are too far away from the actual parameter values. This can happen, for example, if the start value is entered as a positive number but the actual value is negative.
- Forgetting to put NS on the OU card when you have input your own start values.

Summary

This chapter outlined basic information related to using the LISREL computer program for the analysis of CFA models. The focus of the chapter, for the most part, concentrated on particulars related to program input. As such, details of program setup were examined separately for each of the four major input components: problem run specification, data specification, model specification, and output specification. Along the way, hints were provided in an attempt to help the user make the most efficient use of his or her time. An example of the standard LISREL output was provided based on the hypothetical model presented in Figure 1.3. Finally, suggestions for interpreting and preventing error messages were provided.

Section II   Single-Group Analyses

3
Application 1: Validating a Theoretical Construct

Our first application, in broad terms, tests the hypothesis that adolescent self-concept (SC) is a multidimensional construct consisting of four factors: general SC(GSC), academic SC (ASC), English SC (ESC), and mathematics SC (MSC). The theoretical basis for this hypothesis derives from the hierarchical model of SC proposed by Shavelson, Hubner, and Stanton (1976). (For details of the study related to this application, see Byrne & Shavelson, 1986.)

Although a number of studies have supported the multidimensionality of SC, there have been counterarguments that SC is a unidimensional structure. Thus to test the multidimensionality of SC against the counterhypothesis, the primary hypothesis is tested against two alternative hypotheses: that SC is a two-factor structure consisting of an academic component (ASC) and a general component (GSC) and that SC is a unidimensional construct.

We now examine each of these hypotheses separately, and in more detail.

Hypothesis 1: Self-Concept Is a Four-Factor Structure

The model to be tested in Hypothesis 1 postulates a priori that SC is a four-factor structure consisting of GSC, ASC, ESC, and MSC. It is presented schematically in Figure 3.1

To work with LISREL, we must now translate what we see in the model into a set of computer statements that define the CFA model to be tested. Let's begin by dissecting the model presented in Figure 3.1 and listing what we observe.

1. There are four SC factors (ξ_1–ξ_4).
2. The four factors are intercorrelated (ϕs).
3. There are 12 observed measures (Xs).
4. These observed measures load onto the factors in the following pattern: X_1–X_3 load onto Factor 1; X_4–X_6 load onto Factor 2; X_7–X_9 load onto Factor 3; and X_{10}–X_{12} load onto Factor 4.

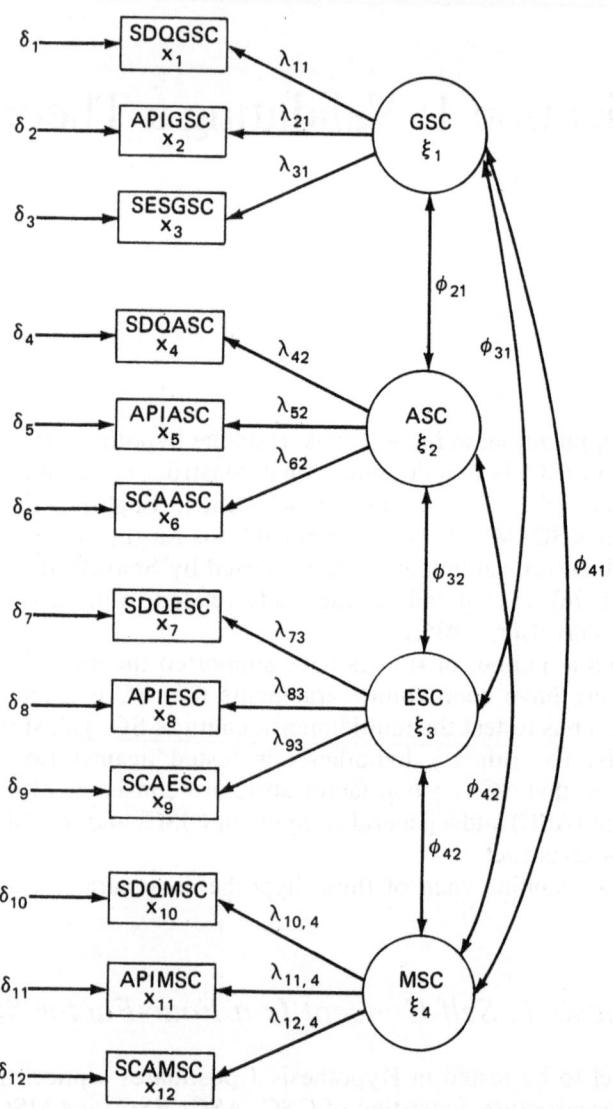

FIGURE 3.1. Hypothesized Structure of Four-Factor Model of Self-Concept.

5. Each X variable loads on one and only one factor.
6. Measurement error is taken into account for each X variable (δs).
7. The errors of measurement are uncorrelated.

Summarizing these observations, we can now present a more formal description of our hypothesized model. As such, we can say that the CFA model presented here hypothesizes a priori that:

(a) SC responses can be explained by four factors: GSC, ASC, ESC, and MSC.
(b) Each subscale measure has a nonzero loading on the SC factor that it was designed to measure (termed a "target loading") and a zero loading on all other factors (termed "nontarget loadings").
(c) The four SC factors, consistent with the theory, are correlated.
(d) Error/uniqueness[1] terms for each of the measures are uncorrelated.

Table 3.1 summarizes the pattern of parameters to be estimated for the factor loading (lambda X; Λ_X), factor variance-covariance (phi; Φ) and error variance-covariance (theta delta; Θ_δ) matrices. The λs, ϕs, and δs represent the parameters to be estimated; the 0s and 1s, the fixed parame-

TABLE 3.1. Pattern of Estimated Parameters for Hypothesized Four-Factor CFA Model

| Factor Loading Matrix (Λ_X) | | GSC | ASC | ESC | MSC |
Measure	X	(ξ_1)	(ξ_2)	(ξ_3)	(ξ_4)
SDQGSC	1	1.00	.0	.0	.0
APIGSC	2	λ_{21}	.0	.0	.0
SESGSC	3	λ_{31}	.0	.0	.0
SDQASC	4	.0	1.00	.0	.0
APIASC	5	.0	λ_{52}	.0	.0
SCAASC	6	.0	λ_{62}	.0	.0
SDQESC	7	.0	.0	1.00	.0
APIESC	8	.0	.0	λ_{83}	.0
SCAESC	9	.0	.0	λ_{93}	.0
SDQMSC	10	.0	.0	.0	1.00
APIMSC	11	.0	.0	.0	$\lambda_{11,4}$
SCAMSC	12	.0	.0	.0	$\lambda_{12,4}$

[1]The term uniqueness is used in the factor analytic sense to mean a composite of random measurement error and specific measurement error associated with a particular measuring instrument; in cross-sectional studies, the two cannot be separated (Gerbing & Anderson, 1984).

TABLE 3.1. Continued

Factor Variance-Covariance Matrix (Φ)

	GSC (ϕ_{11})	ASC (ϕ_{22})	ESC (ϕ_{33})	MSC (ϕ_{44})
GSC	ϕ_{11}			
ASC	ϕ_{21}	ϕ_{22}		
ESC	ϕ_{31}	ϕ_{32}	ϕ_{33}	
MSC	ϕ_{41}	ϕ_{42}	ϕ_{43}	ϕ_{44}

Error Variance-Covariance Matrix (Θ_δ)

	X_1	X_2	X_3	X_4	X_5	X_6	X_7	X_8	X_9	X_{10}	X_{11}	X_{12}
X_1	$\theta_{\delta 11}$	.0	.0	.0	.0	.0	.0	.0	.0	.0	.0	.0
X_2	.0	$\theta_{\delta 22}$	.0	.0	.0	.0	.0	.0	.0	.0	.0	.0
X_3	.0	.0	$\theta_{\delta 33}$	.0	.0	.0	.0	.0	.0	.0	.0	.0
X4	.0	.0	.0	$\theta_{\delta 44}$	.0	.0	.0	.0	.0	.0	.0	.0
X_5	.0	.0	.0	.0	$\theta_{\delta 55}$	.0	.0	.0	.0	.0	.0	.0
X_6	.0	.0	.0	.0	.0	$\theta_{\delta 66}$	.0	.0	.0	.0	.0	.0
X_7	.0	.0	.0	.0	.0	.0	$\theta_{\delta 77}$	.0	.0	.0	.0	.0
X_8	.0	.0	.0	.0	.0	.0	.0	$\theta_{\delta 88}$	.0	.0	.0	.0
X_9	.0	.0	.0	.0	.0	.0	.0	.0	$\theta_{\delta 99}$	.0	.0	.0
X_{10}	.0	.0	.0	.0	.0	.0	.0	.0	.0	$\theta_{\delta 10,10}$	.0	.0
X_{11}	.0	.0	.0	.0	.0	.0	.0	.0	.0	.0	$\theta_{\delta 11,11}$	.0
X_{12}	.0	.0	.0	.0	.0	.0	.0	.0	.0	.0	.0	$\theta_{\delta 12,12}$

ters. For purposes of identification, the first of each congeneric set[2] of SC measures was fixed to 1.0.

[2] A set of measures is congeneric if they all purport to assess the same construct, except for errors of measurement (Jöreskog, 1971b).

1. LISREL Input

Now we are ready to translate this information into LISREL language, which is needed in setting up the input statements describing our postulated model. The basic LISREL input for this model (Model 1), including the data, is presented in Table 3.2. The data are in the form of a correlation matrix.[3]

Although the transition from Table 3.1 to Table 3.2 is fairly straightforward, a few words of explanation are in order.

1. NI = 15—There are 15 instead of 12 input variables because in the original study, the analyses included three achievement measures, labeled as 'GPA,' 'ENG,' and 'MATH.'

2. The SE card was used as a consequence of three changes required in reading the data correlation matrix. These are:

(a) We do not want to include the variables GPA, ENG, and MATH (13, 14, and 15, respectively).

TABLE 3.2. LISREL Input for Model 1

```
CONFIRMATORY FACTOR ANALYSIS INITIAL MODEL
DA NI=15 NO=996 MA=KM
LA
*
'SDQGSC' 'SDQASC' 'SDQESC' 'SDQMSC' 'APIGSC' 'SESGSC' 'APIASC' 'SCAASC'
'APIESC' 'SCAESC' 'APIMSC' 'SCAMSC' 'GPA' 'ENG' 'MATH'
KM SY
(15F4.3)
1000
 3011000
 289 3881000
 170 453 0121000
 630 266 227 2001000
 786 306 299 225 6351000
 522 619 389 346 579 5371000
 216 675 343 472 216 283 5451000
 156 442 705 014 190 190 440 3691000
 128 470 543 069 131 174 396 589 6271000
 177 475 066 864 270 257 426 489 142 0961000
 135 424 027 828 188 187 367 577 028 146 8061000
 010 506 162 395 006 063 374 661 147 375 321 4421000
-008 457 219 236-020 039 326 523 261 541 182 241 7941000
-017 349 057 562 001 034 262 489 039 164 477 624 739 5141000
SELECTION
 1 5 6 2 8 3 9 10 4 11 12/
MO NX=11 NK=4 LX=FU PH=SY TD=DI
FR LX(2,1) LX(3,1) LX(5,2) LX(7,3) LX(8,3) LX(10,4) LX(11,4)
ST 1.0 LX(1,1) LX(4,2) LX(6,3) LX(9,4)
ST .7 LX(2,1) LX(3,1) LX(5,2) LX(7,3) LX(8,3) LX(10,4) LX(11,4)
ST .5 PH(1,1) PH(2,2) PH(3,3) PH(4,4)
ST .2 PH(2,1) PH(3,1) PH(3,2) PH(4,1) PH(4,2) PH(4,3)
ST .3 TD(1,1)-TD(11,11)
OU NS SE TV RS MI
```

[3]Correlations were computed using the SPSS program based on pairwise deletion of missing data. Subsequent to these analyses, however, I have since found good reason to base my analyses on correlation matrices derived from listwise deletion of missing data. I strongly recommend the latter, since it can often eliminate LISREL convergence problems.

(b) The correlation matrix was computed using an ordering of the variables that differed from their order of input (i.e., SDQGSC, APIGSC, SESGSC, SDQASC, APIASC, SCAASC, SDQESC, APIESC, SCAESC, SDQMSC, APIMSC, SCAMSC).

(c) A preliminary exploratory factor analysis of the API revealed the Student Self subscale (measuring ASC) to be problematic; only 10 of the 25 items loaded >0.25 on the ASC factor. Subsequently, this subscale was deleted from the analyses as one measure of ASC. Elimination of the APIASC led to two important alterations to the pattern of esti-

TABLE 3.3. Revised Pattern of Estimated Parameters for Hypothesized Four-Factor CFA Model

Factor Loading Matrix (Λ_x)

Measure	X	GSC (ξ_1)	ASC (ξ_2)	ESC (ξ_3)	MSC (ξ_4)
SDQGSC	1	1.00	.0	.0	.0
APIGSC	2	λ_{21}	.0	.0	.0
SESGSC	3	λ_{31}	.0	.0	.0
SDQASC	4	.0	1.00	.0	.0
SCAASC	5	.0	λ_{52}	.0	.0
SDQESC	6	.0	.0	1.00	.0
APIESC	7	.0	.0	λ_{73}	.0
SCAESC	8	.0	.0	λ_{83}	.0
SDQMSC	9	.0	.0	.0	1.00
APIMSC	10	.0	.0	.0	$\lambda_{10,4}$
SCAMSC	11	.0	.0	.0	$\lambda_{11,4}$

Factor Variance-Covariance Matrix (Φ)

	GSC (ϕ_{11})	ASC (ϕ_{22})	ESC (ϕ_{33})	MSC (ϕ_{44})
GSC	ϕ_{11}			
ASC	ϕ_{21}	ϕ_{22}		
ESC	ϕ_{31}	ϕ_{32}	ϕ_{33}	
MSC	ϕ_{41}	ϕ_{42}	ϕ_{43}	ϕ_{44}

TABLE 3.3. Continued

Error Variance-Covariance Matrix (Θ_δ)

	X_1	X_2	X_3	X_4	X_5	X_6	X_7	X_8	X_9	X_{10}	X_{11}
X_1	$\theta\delta_{11}$	.0	.0	.0	.0	.0	.0	.0	.0	.0	.0
X_2	.0	$\theta\delta_{22}$	.0	.0	.0	.0	.0	.0	.0	.0	.0
X_3	.0	.0	$\theta\delta_{33}$	.0	.0	.0	.0	.0	.0	.0	.0
X_4	.0	.0	.0	$\theta\delta_{44}$	.0	.0	.0	.0	.0	.0	.0
X_5	.0	.0	.0	.0	$\theta\delta_{55}$	.0	.0	.0	.0	.0	.0
X_6	.0	.0	.0	.0	.0	$\theta\delta_{66}$	.0	.0	.0	.0	.0
X_7	.0	.0	.0	.0	.0	.0	$\theta\delta_{77}$	.0	.0	.0	.0
X_8	.0	.0	.0	.0	.0	.0	.0	$\theta\delta_{88}$	.0	.0	.0
X_9	.0	.0	.0	.0	.0	.0	.0	.0	$\theta\delta_{99}$	.0	.0
X_{10}	.0	.0	.0	.0	.0	.0	.0	.0	.0	$\theta\delta_{10,10}$	.0
X_{11}	.0	.0	.0	.0	.0	.0	.0	.0	.0	.0	$\theta\delta_{11,11}$

mated parameters as shown in Table 3.2: (i) the subscripted numbering of the λ parameters representing ASC, ESC, and MSC changed, resulting in the revised pattern of loadings as shown in Table 3.3, and (ii) the number of λ_s and $\Theta_\delta s$ was reduced from 12 to 11.

3. The default values have been used for the MO card. Thus, although the keywords have not been listed, the LX, PH, and TD matrices are specifed as being fixed, free, and free, respectively.

2. LISREL Output

For purposes of discussion, the entire output is presented here for Application 1 only; subsequently, selected segments of printout material will be presented. The discussion will focus on two major aspects of the printout: the LISREL summary of the specified model to be estimated and the assessment of model fit. The printed output for Model 1 is presented in Table 3.4.

TABLE 3.4. LISREL Output for Model 1

L I S R E L   V I

BY

KARL G JORESKOG AND DAG SORBOM

CONFIRMATORY FACTOR ANALYSIS INITIAL MODEL

THE FOLLOWING LISREL CONTROL LINES HAVE BEEN READ :

```
DA NI=15 NO=996 MA=KM
LA
*
'SDQGSC' 'SDQASC' 'SDQESC' 'SDQMSC' 'APIGSC' 'SESGSC' 'APIASC' 'SCAASC'
'APIESC' 'SCAESC' 'APIMSC' 'SCAMSC' 'GPA' 'ENG' 'MATH'
KM SY
(15F4.3)
SELECTION
 1  5  6  2  8  3  9  10  4  11  12/
MO NX=11 NK=4 LX=FU PH=SY TD=SY,FI
FR LX(2,1) LX(3,1) LX(5,2) LX(7,3) LX(8,3) LX(10,4) LX(11,4)
FR TD(1,1) TD(2,2) TD(3,3) TD(4,4) TD(5,5) TD(6,6) TD(7,7)
FR TD(8,8) TD(9,9) TD(10,10) TD(11,11)
ST 1,0 LX(1,1) LX(4,2) LX(6,3) LX(9,4)
ST .7 LX(2,1) LX(3,1) LX(5,2) LX(7,3) LX(8,3) LX(10,4) LX(11,4)
ST .5 PH(1,1) PH(2,2) PH(3,3) PH(4,4)
ST .2 PH(2,1) PH(3,1) PH(3,2) PH(4,1) PH(4,2) PH(4,3)
ST .3 TD(1,1) TD(2,2) TD(3,3) TD(4,4) TD(5,5) TD(6,6) TD(7,7)
ST .3 TD(8,8) TD(9,9) TD(10,10) TD(11,11)
OU NS SE TV RS MI
```

NUMBER OF INPUT VARIABLES 15

NUMBER OF Y - VARIABLES   0

NUMBER OF X - VARIABLES   11

NUMBER OF ETA - VARIABLES   0

NUMBER OF KSI - VARIABLES   4

NUMBER OF OBSERVATIONS   996

OUTPUT REQUESTED

TECHNICAL OUTPUT	NO
STANDARD ERRORS	YES
T - VALUES	YES
CORRELATIONS OF ESTIMATES	NO
FITTED MOMENTS	YES
TOTAL EFFECTS	NO
VARIANCES AND COVARIANCES	NO
MODIFICATION INDICES	YES
FACTOR SCORES REGRESSIONS	NO
FIRST ORDER DERIVATIVES	NO
STANDARDIZED SOLUTION	NO
PARAMETER PLOTS	NO
AUTOMATIC MODIFICATION	NO

TABLE 3.4. Continued

```
CORRELATION MATRIX TO BE ANALYZED
```

	SDQGSC	APIGSC	SESGSC	SDQASC	SCAASC
SDQGSC	1.000				
APIGSC	0.630	1.000			
SESGSC	0.786	0.635	1.000		
SDQASC	0.301	0.266	0.306	1.000	
SCAASC	0.216	0.216	0.283	0.675	1.000
SDQESC	0.289	0.227	0.299	0.388	0.343
APIESC	0.156	0.190	0.190	0.442	0.369
SCAESC	0.128	0.131	0.174	0.470	0.589
SDQMSC	0.170	0.200	0.225	0.453	0.472
APIMSC	0.177	0.270	0.257	0.475	0.489
SCAMSC	0.135	0.188	0.187	0.424	0.577

	SDQESC	APIESC	SCAESC	SDQMSC	APIMSC
	1.000				
	0.705	1.000			
	0.543	0.627	1.000		
	0.012	0.014	0.069	1.000	
	0.066	0.142	0.096	0.864	1.000
	0.027	0.028	0.146	0.828	0.806

```
CORRELATION MATRIX TO BE ANALYZED
```

	SCAMSC
SCAMSC	1.000

```
DETERMINANT = 0.320102D-03
```

PARAMETER SPECIFICATIONS

```
LAMBDA X
```

	KSI 1	KSI 2	KSI 3	KSI 4
SDQGSC	0	0	0	0
APIGSC	1	0	0	0
SESGSC	2	0	0	0
SDQASC	0	0	0	0
SCAASC	0	3	0	0
SDQESC	0	0	0	0
APIESC	0	0	4	0
SCAESC	0	0	5	0
SDQMSC	0	0	0	0
APIMSC	0	0	0	6
SCAMSC	0	0	0	7

```
PHI
```

	KSI 1	KSI 2	KSI 3	KSI 4
KSI 1	8			
KSI 2	9	10		
KSI 3	11	12	13	
KSI 4	14	15	16	17

TABLE 3.4. Continued

THETA DELTA

	SDQGSC	APIGSC	SESGSC	SDQASC	SCAASC
SDQGSC	18				
APIGSC	0	19			
SESGSC	0	0	20		
SDQASC	0	0	0	21	
SCAASC	0	0	0	0	22
SDQESC	0	0	0	0	0
APIESC	0	0	0	0	0
SCAESC	0	0	0	0	0
SDQMSC	0	0	0	0	0
APIMSC	0	0	0	0	0
SCAMSC	0	0	0	0	0

	SDQESC	APIESC	SCAESC	SDQMSC	APIMSC
SDQESC	23				
APIESC	0	24			
SCAESC	0	0	25		
SDQMSC	0	0	0	26	
APIMSC	0	0	0	0	27
SCAMSC	0	0	0	0	0

THETA DELTA

	SCAMSC
SCAMSC	28

STARTING VALUES

LAMBDA X

	KSI 1	KSI 2	KSI 3	KSI 4
SDQGSC	1.000	0.0	0.0	0.0
APIGSC	0.700	0.0	0.0	0.0
SESGSC	0.700	0.0	0.0	0.0
SDQASC	0.0	1.000	0.0	0.0
SCAASC	0.0	0.700	0.0	0.0
SDQESC	0.0	0.0	1.000	0.0
APIESC	0.0	0.0	0.700	0.0
SCAESC	0.0	0.0	0.700	0.0
SDQMSC	0.0	0.0	0.0	1.000
APIMSC	0.0	0.0	0.0	0.700
SCAMSC	0.0	0.0	0.0	0.700

PHI

	KSI 1	KSI 2	KSI 3	KSI 4
KSI 1	0.500			
KSI 2	0.200	0.500		
KSI 3	0.200	0.200	0.500	
KSI 4	0.200	0.200	0.200	0.500

THETA DELTA

	SDQGSC	APIGSC	SESGSC	SDQASC	SCAASC
SDQGSC	0.300				
APIGSC	0.0	0.300			
SESGSC	0.0	0.0	0.300		
SDQASC	0.0	0.0	0.0	0.300	
SCAASC	0.0	0.0	0.0	0.0	0.300
SDQESC	0.0	0.0	0.0	0.0	0.0
APIESC	0.0	0.0	0.0	0.0	0.0
SCAESC	0.0	0.0	0.0	0.0	0.0
SDQMSC	0.0	0.0	0.0	0.0	0.0
APIMSC	0.0	0.0	0.0	0.0	0.0
SCAMSC	0.0	0.0	0.0	0.0	0.0

TABLE 3.4. Continued

	SDQESC	APIESC	SCAESC	SDQMSC	APIMSC

0.300				
0.0	0.300			
0.0	0.0	0.300		
0.0	0.0	0.0	0.300	
0.0	0.0	0.0	0.0	0.300
0.0	0.0	0.0	0.0	0.0

THETA DELTA

	SCAMSC
SCAMSC	0.300

SQUARED MULTIPLE CORRELATIONS FOR X - VARIABLES

SDQGSC	APIGSC	SESGSC	SDQASC	SCAASC
0.700	0.700	0.700	0.700	0.700

SDQESC	APIESC	SCAESC	SDQMSC	APIMSC
0.700	0.700	0.700	0.700	0.700

SQUARED MULTIPLE CORRELATIONS FOR X - VARIABLES

SCAMSC
0.700

TOTAL COEFFICIENT OF DETERMINATION FOR X - VARIABLES
IS 0.997

LISREL ESTIMATES (MAXIMUM LIKELIHOOD)

LAMBDA X

	KSI 1	KSI 2	KSI 3	KSI 4
SDQGSC	1.000	0.0	0.0	0.0
APIGSC	0.819	0.0	0.0	0.0
SESGSC	1.030	0.0	0.0	0.0
SDQASC	0.0	1.000	0.0	0.0
SCAASC	0.0	1.027	0.0	0.0
SDQESC	0.0	0.0	1.000	0.0
APIESC	0.0	0.0	1.075	0.0
SCAESC	0.0	0.0	0.973	0.0
SDQMSC	0.0	0.0	0.0	1.000
APIMSC	0.0	0.0	0.0	0.976
SCAMSC	0.0	0.0	0.0	0.944

PHI

		KSI 1	KSI 2	KSI 3	KSI 4
KSI	1	0.762			
KSI	2	0.271	0.657		
KSI	3	0.200	0.410	0.613	
KSI	4	0.209	0.482	0.061	0.881

TABLE 3.4. Continued

THETA DELTA

	SDQGSC	APIGSC	SESGSC	SDQASC	SCAASC
SDQGSC	0.238				
APIGSC	0.0	0.489			
SESGSC	0.0	0.0	0.192		
SDQASC	0.0	0.0	0.0	0.343	
SCAASC	0.0	0.0	0.0	0.0	0.306
SDQESC	0.0	0.0	0.0	0.0	0.0
APIESC	0.0	0.0	0.0	0.0	0.0
SCAESC	0.0	0.0	0.0	0.0	0.0
SDQMSC	0.0	0.0	0.0	0.0	0.0
APIMSC	0.0	0.0	0.0	0.0	0.0
SCAMSC	0.0	0.0	0.0	0.0	0.0

SDQESC	APIESC	SCAESC	SDQMSC	APIMSC
0.387				
0.0	0.292			
0.0	0.0	0.420		
0.0	0.0	0.0	0.119	
0.0	0.0	0.0	0.0	0.162
0.0	0.0	0.0	0.0	0.0

THETA DELTA

	SCAMSC
SCAMSC	0.216

SQUARED MULTIPLE CORRELATIONS FOR X - VARIABLES

SDQGSC	APIGSC	SESGSC	SDQASC	SCAASC
0.762	0.511	0.808	0.657	0.694

SDQESC	APIESC	SCAESC	SDQMSC	APIMSC
0.613	0.708	0.580	0.881	0.838

SQUARED MULTIPLE CORRELATIONS FOR X - VARIABLES

SCAMSC
0.784

TOTAL COEFFICIENT OF DETERMINATION FOR X - VARIABLES
IS   0.999

MEASURES OF GOODNESS OF FIT FOR THE WHOLE MODEL :

CHI-SQUARE WITH   38 DEGREES OF FREEDOM IS   627.57
(PROB. LEVEL = 0.0   )

GOODNESS OF FIT INDEX IS 0.892

ADJUSTED GOODNESS OF FIT INDEX IS 0.813

ROOT MEAN SQUARE RESIDUAL IS   0.048

TABLE 3.4. Continued

STANDARD ERRORS

LAMBDA X

	KSI 1	KSI 2	KSI 3	KSI 4
SDQGSC	0.0	0.0	0.0	0.0
APIGSC	0.032	0.0	0.0	0.0
SESGSC	0.033	0.0	0.0	0.0
SDQASC	0.0	0.0	0.0	0.0
SCAASC	0.0	0.039	0.0	0.0
SDQESC	0.0	0.0	0.0	0.0
APIESC	0.0	0.0	0.042	0.0
SCAESC	0.0	0.0	0.041	0.0
SDQMSC	0.0	0.0	0.0	0.0
APIMSC	0.0	0.0	0.0	0.020
SCAMSC	0.0	0.0	0.0	0.021

PHI

	KSI 1	KSI 2	KSI 3	KSI 4
KSI 1	0.047			
KSI 2	0.029	0.045		
KSI 3	0.026	0.031	0.045	
KSI 4	0.029	0.033	0.026	0.045

THETA DELTA

	SDQGSC	APIGSC	SESGSC	SDQASC	SCAASC
SDQGSC	0.021				
APIGSC	0.0	0.025			
SESGSC	0.0	0.0	0.020		
SDQASC	0.0	0.0	0.0	0.023	
SCAASC	0.0	0.0	0.0	0.0	0.023
SDQESC	0.0	0.0	0.0	0.0	0.0
APIESC	0.0	0.0	0.0	0.0	0.0
SCAESC	0.0	0.0	0.0	0.0	0.0
SDQMSC	0.0	0.0	0.0	0.0	0.0
APIMSC	0.0	0.0	0.0	0.0	0.0
SCAMSC	0.0	0.0	0.0	0.0	0.0

	SDQESC	APIESC	SCAESC	SDQMSC	APIMSC
	0.024				
	0.0	0.023			
	0.0	0.0	0.025		
	0.0	0.0	0.0	0.010	
	0.0	0.0	0.0	0.0	0.011
	0.0	0.0	0.0	0.0	0.0

THETA DELTA

	SCAMSC
SCAMSC	0.013

T-VALUES

LAMBDA X

	KSI 1	KSI 2	KSI 3	KSI 4
SDQGSC	0.0	0.0	0.0	0.0
APIGSC	25.325	0.0	0.0	0.0
SESGSC	31.477	0.0	0.0	0.0
SDQASC	0.0	0.0	0.0	0.0
SCAASC	0.0	26.684	0.0	0.0
SDQESC	0.0	0.0	0.0	0.0
APIESC	0.0	0.0	25.555	0.0
SCAESC	0.0	0.0	23.650	0.0
SDQMSC	0.0	0.0	0.0	0.0
APIMSC	0.0	0.0	0.0	49.610
SCAMSC	0.0	0.0	0.0	45.571

TABLE 3.4. Continued

PHI

		KSI 1	KSI 2	KSI 3	KSI 4
KSI	1	16.228			
KSI	2	9.424	14.469		
KSI	3	7.552	13.392	13.754	
KSI	4	7.117	14.440	2.305	19.410

THETA DELTA

	SDQGSC	APIGSC	SESGSC	SDQASC	SCAASC
SDQGSC	11.583				
APIGSC	0.0	19.425			
SESGSC	0.0	0.0	9.367		
SDQASC	0.0	0.0	0.0	14.977	
SCAASC	0.0	0.0	0.0	0.0	13.574
SDQESC	0.0	0.0	0.0	0.0	0.0
APIESC	0.0	0.0	0.0	0.0	0.0
SCAESC	0.0	0.0	0.0	0.0	0.0
SDQMSC	0.0	0.0	0.0	0.0	0.0
APIMSC	0.0	0.0	0.0	0.0	0.0
SCAMSC	0.0	0.0	0.0	0.0	0.0

SDQESC	APIESC	SCAESC	SDQMSC	APIMSC
16.112				
0.0	12.931			
0.0	0.0	16.952		
0.0	0.0	0.0	11.576	
0.0	0.0	0.0	0.0	14.498
0.0	0.0	0.0	0.0	0.0

THETA DELTA

	SCAMSC
SCAMSC	17.070

FITTED MOMENTS AND RESIDUALS

FITTED MOMENTS

	SDQGSC	APIGSC	SESGSC	SDQASC	SCAASC
SDQGSC	1.000				
APIGSC	0.624	1.000			
SESGSC	0.785	0.643	1.000		
SDQASC	0.271	0.222	0.279	1.000	
SCAASC	0.278	0.228	0.287	0.675	1.000
SDQESC	0.200	0.164	0.206	0.410	0.421
APIESC	0.215	0.176	0.221	0.441	0.453
SCAESC	0.195	0.159	0.200	0.399	0.410
SDQMSC	0.209	0.171	0.215	0.482	0.496
APIMSC	0.204	0.167	0.210	0.471	0.484
SCAMSC	0.197	0.162	0.203	0.455	0.468

SDQESC	APIESC	SCAESC	SDQMSC	APIMSC
1.000				
0.659	1.000			
0.596	0.641	1.000		
0.061	0.065	0.059	1.000	
0.059	0.063	0.057	0.859	1.000
0.057	0.061	0.056	0.831	0.811

TABLE 3.4. Continued

FITTED MOMENTS

	SCAMSC
SCAMSC	1.000

FITTED RESIDUALS

	SDQGSC	APIGSC	SESGSC	SDQASC	SCAASC
SDQGSC	-0.000				
APIGSC	0.006	-0.000			
SESGSC	0.001	-0.008	-0.000		
SDQASC	0.030	0.044	0.027	-0.000	
SCAASC	-0.062	-0.012	-0.004	-0.000	-0.000
SDQESC	0.089	0.063	0.093	-0.022	-0.078
APIESC	-0.059	0.014	-0.031	0.001	-0.084
SCAESC	-0.067	-0.028	-0.026	0.071	0.179
SDQMSC	-0.039	0.029	0.010	-0.029	-0.024
APIMSC	-0.027	0.103	0.047	0.004	0.005
SCAMSC	-0.062	0.026	-0.016	-0.031	0.109

	SDQESC	APIESC	SCAESC	SDQMSC	APIMSC
	0.000				
	0.046	-0.000			
	-0.053	-0.014	-0.000		
	-0.049	-0.051	0.010	-0.000	
	0.007	0.079	0.039	0.005	-0.000
	-0.030	-0.033	0.090	-0.003	-0.005

FITTED RESIDUALS

	SCAMSC
SCAMSC	-0.000

NORMALIZED RESIDUALS

	SDQGSC	APIGSC	SESGSC	SDQASC	SCAASC
SDQGSC	-0.000				
APIGSC	0.155	-0.000			
SESGSC	0.033	-0.209	-0.000		
SDQASC	0.918	1.358	0.822	-0.000	
SCAASC	-1.892	-0.368	-0.109	-0.000	-0.000
SDQESC	2.756	1.968	2.877	-0.645	-2.278
APIESC	-1.817	0.433	-0.965	0.032	-2.413
SCAESC	-2.060	-0.884	-0.815	2.078	5.223
SDQMSC	-1.207	0.892	0.297	-0.838	-0.670
APIMSC	-0.834	3.201	1.448	0.123	0.154
SCAMSC	-1.929	0.820	-0.502	-0.900	3.119

	SDQESC	APIESC	SCAESC	SDQMSC	APIMSC
	0.000				
	1.217	-0.000			
	-1.444	-0.374	-0.000		
	-1.527	-1.607	0.319	-0.000	
	0.220	2.473	1.214	0.115	-0.000
	-0.948	-1.051	2.848	-0.077	-0.118

NORMALIZED RESIDUALS

	SCAMSC
SCAMSC	-0.000

TABLE 3.4. Continued

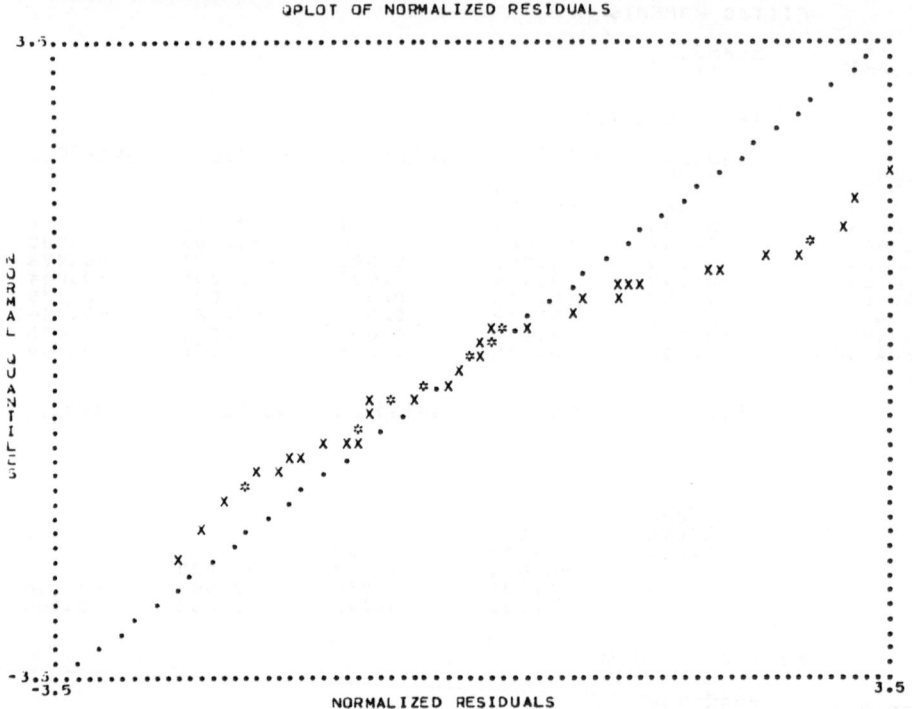

THE PROBLEM REQUIRED 1475 DOUBLE PRECISION WORDS. THE CPU-TIME WAS   1.61 SECONDS

MODIFICATION INDICES

LAMBDA X

	KSI 1	KSI 2	KSI 3	KSI 4
SDQGSC	0.0	8.190	2.212	13.456
APIGSC	0.0	2.504	0.616	7.137
SESGSC	0.0	2.870	0.838	3.127
SDQASC	7.567	0.0	3.019	4.608
SCAASC	7.568	0.0	3.019	4.608
SDQESC	35.455	12.289	0.0	3.338
APIESC	10.685	15.040	0.0	0.569
SCAESC	6.240	63.076	0.0	7.636
SDQMSC	0.978	25.956	25.086	0.0
APIMSC	9.510	6.014	21.747	0.0
SCAMSC	5.121	10.652	0.578	0.0

PHI

	KSI 1	KSI 2	KSI 3	KSI 4
KSI 1	0.0			
KSI 2	0.0	0.0		
KSI 3	0.0	0.0	0.0	
KSI 4	0.0	0.0	0.0	0.0

TABLE 3.4. Continued

THETA DELTA

	SDQGSC__	APIGSC__	SESGSC__	SDQASC__	SCAASC__
SDQGSC	0.0				
APIGSC	3.228	0.0			
SESGSC	3.576	10.163	0.0		
SDQASC	16.220	0.002	5.257	0.0	
SCAASC	7.077	2.533	4.447	0.0	0.0
SDQESC	13.057	0.812	2.309	0.210	24.709
APIESC	6.473	9.387	1.189	4.491	67.802
SCAESC	7.846	0.841	0.261	2.360	199.760
SDQMSC	2.785	6.962	0.205	6.409	20.394
APIMSC	12.626	23.825	2.883	2.206	24.775
SCAMSC	0.306	0.002	2.173	33.055	142.218

SDQESC__	APIESC__	SCAESC__	SDQMSC__	APIMSC__
0.0				
125.156	0.0			
60.026	9.044	0.0		
3.117	9.220	1.468	0.0	
0.076	109.267	39.396	26.184	0.0
2.472	32.724	29.814	5.016	5.454

THETA DELTA

	SCAMSC__
SCAMSC	0.0

MAXIMUM MODIFICATION INDEX IS 199.76 FOR ELEMENT (8, 5)
OF THETA DELTA

2.1. Specified LISREL Model

There are several advantages in having this information printed in the output. First, it enables the user to check for syntax errors in the input of the control cards and model specifications; this can be helpful in solving problems associated with a problematic computer run. Second, it provides a double check on the starting values of both the fixed and free parameters in the model. Finally, the numbering of estimated parameters acts as a countercheck that the fixed and free parameters are correct. This information also enables the user to confirm that the number of degrees of freedom provided by the program is accurate.

The number of degrees of freedom is equal to the difference between the number of parameters being estimated (Σ; the restricted correlation matrix) and the total number of parameters in the model (S; the sample correlation matrix). The total number of parameters in the model equals $p(p+1)/2$ where p = observed variable.

With respect to the present model, there are 11 observed variables; thus there are 66 ([11 × 12]/2) parameters in the entire model. Turning to page 46 of the printout, we can see that there are 28 parameters to be estimated. This means that in fitting the restricted (i.e., hypothesized) model ($\hat{\Sigma}$) to the sample data (S), we should have 38 degrees of freedom.

2.2. Assessment of Model Fit

The most important issue associated with the analysis of LISREL models is the assessment of fit between the hypothesized model and the sample data. If the goodness-of-fit is inadequate, the next logical step is to detect the source of misfit in the model.

Many factors are taken into account in assessing the adequacy of a hypothesized model. Let's now examine the major aspects of this issue.

2.2.1. Feasibility of Parameter Estimates. The first step in examining model fit is to determine whether the parameter estimates are reasonable. If some parameters fall outside the admissable range, this is a clear indication that either the model is wrong or the input matrix lacks sufficient information. Examples of parameter estimates that are considered to be unreasonable are: negative variances, correlations >1.00, and covariance or correlation matrices that are not positive definite. Given any of these conditions, LISREL prints a warning message. Other indicators of a bad model are standard errors that are excessively large and parameter estimates that are highly correlated.

An examination of the printout in Table 3.3 reveals all LISREL estimates to be reasonable. The standard errors range from 0.010 to 0.047, which is highly acceptable.

2.2.2. Adequacy of the Measurement Model. The second step in assessing model fit is to examine the squared multiple correlation (R^2) for each observed variable and the coefficient of determination for all the observed variables jointly. These values should range from zero to 1.00; values close to 1.00 represent good models. Negative values are a clear indication that something is wrong with the postulated model.

The R^2 is an indication of the reliability of each observed measure with respect to its underlying latent construct. In examining the observed measures of GSC, we see that SESGSC was the most reliable ($R^2 = 0.81$), while the APIGSC was the least reliable ($R^2 = 0.51$).

The coefficient of determination is an indication of how well the observed variables, in combination, serve as measuring instruments for all the latent variables jointly; it is a generalized indicator of reliability for the entire measurement model. Looking at the output, we see that the coefficient of determination is remarkably high (0.999), indicating that the measurement model is excellent.

2.2.3. Goodness-of-Fit of the Overall Model. LISREL provides four indices of fit for the model as a whole: χ^2 with its associated degrees of freedom and probability level (ML and GLS only), the goodness-of-fit index (GFI), the adjusted goodness-of-fit index (AGFI), and the root-mean-square residual (RMR). (For a review and assessment of these indices, see Marsh, Balla, & McDonald, 1988).

When the sample size is sufficiently large, χ^2 is a likelihood ratio test statistic that can be used to test the fit between the restricted hypothesized model and the unrestricted sample data. According to this index, as shown in Table 3.3, the overall fit for the initial model is poor (χ^2 (38) = 627.57).

The GFI indicates the relative amount of variance and covariance jointly explained by the model; the AGFI differs from the GFI only in the fact that it adjusts for the number of degrees of freedom in the model. Both indices range from zero to 1.00, with a value close to 1.00 indicating a good fit. Although Jöreskog and Sorböm (1985) argue that the GFI (and AGFI), unlike χ^2, are independent of sample size and robust to departures from multinormality, others have disputed this claim (see, e.g., Marsh, Balla, & McDonald, 1988).

The GFI in our present application is found to be 0.892, thus representing a fairly good fit between the hypothesized model and the observed data. However, when the degrees of freedom are taken into account, the goodness of fit diminishes somewhat (AGFI = 0.813).

The RMR indicates the average discrepancy between the elements in the sample and hypothesized covariance matrices; values range from zero to 1.00. Given a good fit beteen the two models, the RMR will be small; this value should be <0.05. The reader is cautioned, however, that wrong models can also have RMRs <0.05. Thus, it is important not to rely too heavily on this single piece of information in determining model fit.

2.2.4. Subjective Goodness-of-Fit Indices for Overall Model. The sensitivity of the χ^2 likelihood ratio test to sample size, as well as to the violation of various model assumptions (linearity, multinormality, additivity) is now widely known. As an alternative to χ^2, other goodness-of-fit indices have been proprosed (for a review, see Marsh, Balla, & McDonald, 1988). Two of the more commonly used subjective indices are the χ^2/df ratio and the Bentler and Bonett (1980) normed index (BBI).

A variety of acceptable values for the χ^2/df ratio have been proposed, ranging from a low of <1.50 for a sample size of 1000 (Muthén, personal communication, January 1987), through <3.00 (Carmines & McIver, 1981), to <5.00 (Wheaton, Muthén, Alwin, & Summers, 1977). At this point in our knowledge of LISREL models and the χ^2 test statistic, it seems clear that a χ^2/df ratio >2.00 represents an inadequate fit.

The χ^2/df ratio for the hypothesized model in our current application is shown to be 627.57/38 = 16.52. This value clearly represents an unacceptable fit to the observed data.

The BBI ranges from zero to 1.00 and is derived from the comparison of some specified (i.e., restricted) model, with a null model (i.e., one that posits complete independence of all observed measurements). Thus, it

TABLE 3.5. LISREL Input for Null Model

```
MO  NX=11  NK=11  LX=ID  PH=SY,FI  TD=ZE
FR  PH(1,1)  PH(2,2)  PH(3,3)  PH(4,4)  PH(5,5)  PH(6,6)  PH(7,7)
FR  PH(8,8)  PH(9,9)  PH(10,10)  PH(11,11)
ST  .5  PH(1,1)  PH(2,2)  PH(3,3)  PH(4,4)  PH(5,5)  PH(6,6)  PH(7,7)
ST  .5  PH(8,8)  PH(9,9)  PH(10,10)  PH(11,11)
OU  NS  SE
```

provides a measure of complete covariation in the data; a value >0.90 indicates a psychometrically acceptable fit to the observed data.

Related to the present application, the BBI is determined by comparing the fit of our four-factor hypothesized model with that of an 11-factor null model. The null model, then, simply represents one in which each observed variable is hypothesized as measuring one independent factor. The LISREL specification input for the null model is presented in Table 3.5.

The computation of the BBI is $(F_0 - F_1)/F_0$ where F_0 = the χ^2 value of the null model and F_1 = the χ^2 of the restricted model. The χ^2 likelihood ratio for the null model in Application 1 was 7523.68 (55). Thus, the BBI was computed to be: $(7523.68 - 627.57)/7523.68 = 0.917$, which, although it falls within the acceptable range for goodness-of-fit, is only marginally so, thus indicating some degree of misfit in the model.

It is important to emphasize that the χ^2, GFI, AGFI, RMR, χ^2/df, and BBI are measures of overall fit; they *do not* pinpoint areas of misfit in a particular model. To determine this information, it is necessary to examine indices that relate to specific parameters in the model. We turn now to this portion of the output.

2.2.5. Goodness-of-Fit of Individual Model Parameters. LISREL provides several indices that can assist the researcher in isolating parameters that may be contributing to the overall misfit of a hypothesized model. However, these indices provide a statistical approach to the problem only and must be considered in conjunction with the substantive meaningfulness of the model. While the program provides many different indices, only those pertinent to the present application will be described here.

(a) *T-Values.* One of the initial things to look at when searching for misfit in a model is to examine the statistical significance of each parameter. Nonsignificant parameters can be considered unimportant to the model and can be subsequently fixed to a value of 0.0; they are thereby deleted from the model. The statistical significance of parameters can be determined by examining the *t*-values provided by LISREL. These values represent the parameter estimate divided by its standard error. As such, *t*-values provide evidence of whether or not a parameter is significantly different from zero; values >2.00 are generally considered to be statistically significant.

An examination of the current output shows *t*-values ranging from 2.305 to 49.610. All parameters may therefore be considered statistically significant and thus important to the hypothesized model.

(b) *Normalized Residuals*.[4] LISREL provides information on the residual of fit for each parameter, that is, the discrepancy of fit between the sample and hypothesized covariance matrices. These residual covariances are reported in both their real metric and standardized form. The latter are referred to in the program as normalized residuals; they are the easier of the two to interpret since they can be considered analogous to Z-scores. Normalized residuals represent estimates of the number of standard deviations the observed residuals are from the zero residuals that would exist if the model were a perfectly fitting one. Values >2.00 for any element provide a clue as to possible model misspecification; in our output, we see 11 such values.

To assist the researcher in further evaluating model fit, LISREL provides a Q-plot, which graphs the normalized residuals. Residuals that follow the dotted line rising at a 45-degree angle in the Q-plot are indicative of a well-fitting model. Those that deviate widely from the 45-degree line in a nonlinear fashion indicate that the model is in some way misspecified. Boomsma (1982) has noted that such departures from normality tend to be larger for uniquenesses than for other model parameters. Examination of the Q-plot for our SC data shows a clear deviation from the upper portion of the dotted line. Thus, we have further evidence to suggest that certain parameters in the model are misspecified.

(c) *Modification Indices*. For each *fixed* parameter in a specified model, LISREL provides a modification index (MI). This value represents the expected drop in χ^2 if a particular parameter were freely estimated. As such, in a respecification and reestimation of the model, the decrease in χ^2 should be at least equal to the MI; it may, however, be much larger. MIs can therefore be examined in relation to χ^2 with one degree of freedom. All free parameters automatically have MI values equal to zero.

LISREL automatically prints out the fixed parameter having the largest MI. If the researcher is unhappy with the overall fit of the hypothesized model, he or she can respecify a model in which this parameter is set free; the model is then reestimated. It must be emphasized, however, that the decision of whether or not to free this parameter must make substantive sense; if it does not, then consideration can be given to freeing the fixed parameter having the next highest MI. It is important, however, to relax only one parameter at a time.

Although the LISREL VI program provides an option for automatic model respecification based on the MIs, this is never a wise option and is definitely not recommended. Only the researcher is capable of judging the balance between statistical and substantive model fit. Thus, model respecification must remain the decision of the researcher and not of the LISREL program.

---

[4]In the most recent version of LISREL (LISREL VII) the term "standardized residuals" is used.

The process of respecification and reestimation based on the examination of MIs can be repeated until an acceptable model fit is attained. However, the analyses then fall into the category of post hoc analyses and, thus, the researcher must realize that the analyses are then exploratory, rather than confirmatory in character; confirmatory factor analyses ceased once the hypothesized model was rejected due to a poor fit with the observed data. The issue of post hoc model fitting is addressed later in the chapter.

Reviewing the output in Table 3.4, we see that the largest MI is 199.76 for TD(8,5).[5] This parameter represents a covariance between the ESC and ASC subscales of the SCA. Such correlated errors can be substantively meaningful in reflecting minor, possibly sample-specific data covariation not explained by the model (Gerbing & Anderson, 1984; Tanaka & Huba, 1984). Frequently, this covariation results from nonrandom error introduced by a particular measurement method; one example is that of method effects due to the item format associated with subscales of a particular measuring instrument.

Confronted with these results, then, it was considered prudent to take an exploratory approach in establishing a well-fitting model. For purposes of demonstrating these post hoc analyses, we will now continue to fit our four-factor model until we are satisfied that we have reached one that is statistically best fitting, yet substantively meaningful; thus a series of nested alternative models were specified and estimated.

To begin, let us again look at the MIs in Table 3.4. We see that the largest MI for Model 1 represents an error covariance between the English and academic SC subscales of the SCA; we therefore specify a model in which TD(8,5) is set free; we'll call this Model 2. Since the parameter TD(8,5) represents an off-diagonal value, this means that we can no longer specify the TD matrix to be diagonal. Theta delta must now be specified as a symmetric matrix that is fixed (SY,FI). This means, also, that we must specify all diagonal values (variances) to be free as well (TD1,1–TD11,11). These specification changes are illustrated in Table 3.6.

The estimation of Model 2 yielded a χ^2 (37) value of 427.01. The difference in fit between Models 1 and 2, expressed as $\Delta\chi^2$, is 200.56. Since $\Delta\chi^2$ is distributed as χ^2 with degrees of freedom equal to the difference in degrees of freedom between the two models, the significance of this value

---

[5]Strangely, given the same model specifications as shown in Table 3.2, the MIs are accorded different values when the LISREL VI program is used. This is because only the diagonal elements in the TD matrix are considered (the off-diagonals are not taken into consideration). However, if the error-covariance matrix is specified as TD = SY,FI, with all the diagonal elements (TD1,1–TD11,11) specified as free parameters (which represents an equivalent specification), the MIs are identical to those estimated by the LISREL V version when the TD matrix is specified as a diagonal matrix (TD = DI).

TABLE 3.6. LISREL Input for Model 2

```
MO  NX=11  NK=4  LX=FU  PH=SY  TD=SY,FI
FR  LX(2,1)  LX(3,1)  LX(5,2)  LX(7,3)  LX(8,3)  LX(10,4)  LX(11,4)
FR  TD(1,1)  TD(2,2)  TD(3,3)  TD(4,4)  TD(5,5)  TD(6,6)  TD(7,7)
FR  TD(8,8)  TD(9,9)  TD(10,10)  TD(11,11)
FR  TD(9,5)
ST  1.0  LX(1,1)  LX(4,2)  LX(5,3)  LX(9,4)
ST  .7  LX(2,1)  LX(3,1)  LX(5,2)  LX(7,3)  LX(8,3)  LX(10,4)  LX(11,4)
ST  .5  PH(1,1)  PH(2,2)  PH(3,3)  PH(4,4)
ST  .2  PH(2,1)  PH(3,1)  PH(3,2)  PH(4,1)  PH(4,2)  PH(4,3)
ST  .3  TD(1,1)  TD(2,2)  TD(3,3)  TD(4,4)  TD(5,5)  TD(6,6)  TD(7,7)
ST  .3  TD(8,8)  TD(9,9)  TD(10,10)  TD(11,11)
ST  .1  TD(9,5)
OU  NS  SE  TV  RS  MI  SS
```

can be tested statistically; of course, $\Delta\chi^2$ (1) = 200.56 is highly significant and indicates a substantial improvement in model fit. Nonetheless, given a χ^2 (37) = 427.01, we must conclude that there still remains a high degree of misfit in the model.

TABLE 3.7. Summary of Respecification Steps in the Model-fitting Process

Competing models	X^2	df	ΔX^2	Δdf	X^2/df	BBI
0 Null model	7,523.68	55	—	—	—	—
1 Four-factor model	627.57	38	—	—	16.52	.917
2 Model 1 with δ_{85} free	427.01	37	200.56	1	11.54	.943
3 Model 1 with δ_{85} $\delta_{11,5}$ free	322.65	36	122.09	1	8.96	.957
4 Model 1 with δ_{85} $\delta_{11,5}$ $\delta_{10,7}$ free	224.01	35	101.92	1	6.40	.970
5 Model 1 with λ_{61} free δ_{85} $\delta_{11,5}$ $\delta_{10,7}$ free	178.62	34	45.39	1	5.25	.976
6 Model 1 with λ_{61} free δ_{85} $\delta_{11,5}$ $\delta_{10,7}$ $\delta_{11,8}$ free	147.39	33	31.23	1	4.47	.980
7 Model 1 with λ_{61} free δ_{85} $\delta_{11,5}$ $\delta_{10,7}$ $\delta_{11,8}$ $\delta_{10,2}$ free	131.19	32	16.20	1	4.10	.983
8 Model 1 with λ_{61} free δ_{85} $\delta_{11,5}$ $\delta_{10,7}$ $\delta_{11,8}$ $\delta_{10,2}$ δ_{41} free	118.55	31	12.64	1	3.82	.984
9 Model 1 with λ_{61} λ_{82} free δ_{85} $\delta_{11,5}$ $\delta_{10,7}$ $\delta_{11,8}$ $\delta_{10,2}$ δ_{41} free	93.24	30	25.31	1	3.11	.988
10 Model 1 with λ_{61} λ_{82} free δ_{85} $\delta_{11,5}$ $\delta_{10,7}$ $\delta_{11,8}$ $\delta_{10,2}$ δ_{41} δ_{72} free	80.76	29	12.48	1	2.78	.989
11 Model 1 with λ_{61} λ_{82} λ_{12} free δ_{85} $\delta_{11,5}$ $\delta_{10,7}$ $\delta_{11,8}$ $\delta_{10,2}$ δ_{41} δ_{72} free	67.44	28	13.32	1	2.41	.991
12 Model 1 with λ_{61} λ_{82} λ_{12} λ_{93} free δ_{85} $\delta_{11,5}$ $\delta_{10,7}$ $\delta_{11,8}$ $\delta_{10,2}$ δ_{41} δ_{72} free	59.19	27	8.25	1	2.19	.991

TABLE 3.8. LISREL Input for Final Model

```
MO NX=11  NK=4  LX=FU  PH=SY  TD=SY,FI
FR LX(2,1)  LX(3,1)  LX(5,2)  LX(7,3)  LX(8,3)  LX(10,4)  LX(11,4)
FR LX(6,1)  LX(8,2)  LX(1,2)  LX(9,3)
FR TD(1,1)  TD(2,2)  TD(3,3)  TD(4,4)  TD(5,5)  TD(6,6)  TD(7,7)
FR TD(8,8)  TD(9,9)  TD(10,10)  TD(11,11)
FR TD(8,5)  TD(11,5)  TD(10,7)  TD(11,8)  TD(10,2)  TD(4,1)  TD(7,2)
ST 1.0 LX(1,1)  LX(4,2)  LX(6,3)  LX(9,4)
ST .7 LX(2,1)  LX(3,1)  LX(5,2)  LX(7,3)  LX(8,3)  LX(10,4)  LX(11,4)
ST .2 LX(5,1)  LX(8,2)  LX(1,2)  LX(9,3)
ST .5 PH(1,1)  PH(2,2)  PH(3,3)  PH(4,4)
ST .2 PH(2,1)  PH(3,1)  PH(3,2)  PH(4,1)  PH(4,2)  PH(4,3)
ST .3 TD(1,1)  TD(2,2)  TD(3,3)  TD(4,4)  TD(5,5)  TD(6,6)  TD(7,7)
ST .3 TD(8,8)  TD(9,9)  TD(10,10)  TD(11,11)
ST .1 TD(8,5)  TD(11,5)  TD(10,7)  TD(11,8)  TD(10,2)  TD(4,1)  TD(7,2)
OU NS  SE  TV  RS  MI
```

In like fashion, building each time on the assessment of individual estimated parameter fit for each previously specified model, a series of models was subsequently specified and estimated until one was obtained that both exhibited an acceptable statistical fit and indicated a substantively meaningful representation of the observed data. In total, 11 post hoc models were eventually hypothesized before an acceptable fit was obtained. Steps in the model-fitting process in reaching this final model are summarized in Table 3.7.

The χ^2 overall fit of this final model was 59.19 with 27 degrees of freedom ($\chi^2/df = 2.19$; BBI = .991); the input for this final model is shown in Table 3.8, and selected output is presented in Table 3.9; included are the LISREL estimates, goodness-of-fit for the whole model, standard errors, Q-plot, and the MIs. A schematic representation of the final model is shown in Figure 3.2.

TABLE 3.9. Selected LISREL Output for Final Model

LISREL ESTIMATES (MAXIMUM LIKELIHOOD)

LAMBDA X

	KSI 1	KSI 2	KSI 3	KSI 4
SDQGSC	1.000	-0.116	0.0	0.0
APIGSC	0.757	0.0	0.0	0.0
SESGSC	0.955	0.0	0.0	0.0
SDQASC	0.0	1.000	0.0	0.0
SCAASC	0.0	0.971	0.0	0.0
SDQESC	0.194	0.0	1.000	0.0
APIESC	0.0	0.0	1.213	0.0
SCAESC	0.0	0.218	0.854	0.0
SDQMSC	0.0	0.0	-0.063	1.000
APIMSC	0.0	0.0	0.0	0.963
SCAMSC	0.0	0.0	0.0	0.926

PHI

	KSI 1	KSI 2	KSI 3	KSI 4
KSI 1	0.862			
KSI 2	0.322	0.699		
KSI 3	0.151	0.334	0.541	
KSI 4	0.244	0.498	0.025	0.888

TABLE 3.9. Continued

THETA DELTA

	SDQGSC	APIGSC	SESGSC	SDQASC	SCAASC
SDQGSC	0.203				
APIGSC	0.0	0.496			
SESGSC	0.0	0.0	0.214		
SDQASC	0.060	0.0	0.0	0.301	
SCAASC	0.0	0.0	0.0	0.0	0.355
SDQESC	0.0	0.0	0.0	0.0	0.0
APIESC	0.0	0.049	0.0	0.0	0.0
SCAESC	0.0	0.0	0.0	0.0	0.190
SDQMSC	0.0	0.0	0.0	0.0	0.0
APIMSC	0.0	0.052	0.0	0.0	0.0
SCAMSC	0.0	0.0	0.0	0.0	0.132

	SDQESC	APIESC	SCAESC	SDQMSC	APIMSC
	0.368				
	0.0	0.198			
	0.0	0.0	0.471		
	0.0	0.0	0.0	0.113	
	0.0	0.084	0.0	0.0	0.155
	0.0	0.0	0.067	0.0	0.0

THETA DELTA

	SCAMSC
SCAMSC	0.231

SQUARED MULTIPLE CORRELATIONS FOR X - VARIABLES

SDQGSC	APIGSC	SESGSC	SDQASC	SCAASC
0.797	0.504	0.788	0.699	0.645

SDQESC	APIESC	SCAESC	SDQMSC	APIMSC
0.632	0.802	0.529	0.887	0.845

SQUARED MULTIPLE CORRELATIONS FOR X - VARIABLES

SCAMSC
0.769

TOTAL COEFFICIENT OF DETERMINATION FOR X - VARIABLES IS   1.000

MEASURES OF GOODNESS OF FIT FOR THE WHOLE MODEL :

CHI-SQUARE WITH 27 DEGREES OF FREEDOM IS   59.19
(PROB. LEVEL = 0.000)

GOODNESS OF FIT INDEX IS 0.989

ADJUSTED GOODNESS OF FIT INDEX IS 0.974

ROOT MEAN SQUARE RESIDUAL IS   0.021

TABLE 3.9. Continued

STANDARD ERRORS

LAMBDA X

	KSI 1	KSI 2	KSI 3	KSI 4
SDQGSC	0.0	0.033	0.0	0.0
APIGSC	0.032	0.0	0.0	0.0
SESGSC	0.034	0.0	0.0	0.0
SDQASC	0.0	0.0	0.0	0.0
SCAASC	0.0	0.038	0.0	0.0
SDQESC	0.026	0.0	0.0	0.0
APIESC	0.0	0.0	0.050	0.0
SCAESC	0.0	0.039	0.045	0.0
SDQMSC	0.0	0.0	0.022	0.0
APIMSC	0.0	0.0	0.0	0.018
SCAMSC	0.0	0.0	0.0	0.020

PHI

	KSI 1	KSI 2	KSI 3	KSI 4
KSI 1	0.058			
KSI 2	0.035	0.047		
KSI 3	0.027	0.028	0.042	
KSI 4	0.033	0.034	0.025	0.045

THETA DELTA

	SDQGSC	APIGSC	SESGSC	SDQASC	SCAASC
SDQGSC	0.022				
APIGSC	0.0	0.025			
SESGSC	0.0	0.0	0.020		
SDQASC	0.014	0.0	0.0	0.024	
SCAASC	0.0	0.0	0.0	0.0	0.025
SDQESC	0.0	0.0	0.0	0.0	0.0
APIESC	0.0	0.015	0.0	0.0	0.0
SCAESC	0.0	0.0	0.0	0.0	0.019
SDQMSC	0.0	0.0	0.0	0.0	0.0
APIMSC	0.0	0.011	0.0	0.0	0.0
SCAMSC	0.0	0.0	0.0	0.0	0.013

SDQESC	APIESC	SCAESC	SDQMSC	APIMSC
0.023				
0.0	0.024			
0.0	0.0	0.025		
0.0	0.0	0.0	0.010	
0.0	0.010	0.0	0.0	0.011
0.0	0.0	0.013	0.0	0.0

THETA DELTA

	SCAMSC
SCAMSC	0.013

TABLE 3.9. Continued

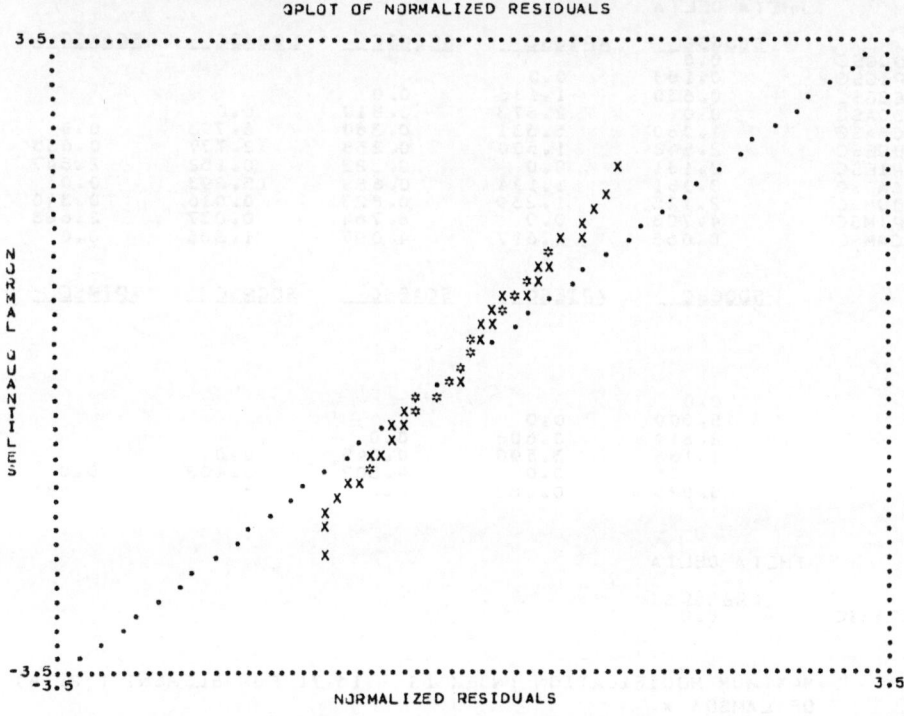

QPLOT OF NORMALIZED RESIDUALS

THE PROBLEM REQUIRED 1915 DOUBLE PRECISION WORDS. THE CPU-TIME WAS   1.86 SECONDS

MODIFICATION INDICES

LAMBDA X

	KSI 1	KSI 2	KSI 3	KSI 4
SDQGSC	0.0	0.0	0.724	1.648
APIGSC	0.0	0.006	0.469	1.568
SESGSC	0.0	0.108	0.127	0.278
SDQASC	0.009	0.0	6.525	6.976
SCAASC	0.020	0.0	5.672	9.374
SDQESC	0.0	3.138	0.0	1.915
APIESC	0.788	2.918	0.0	9.817
SCAESC	0.837	0.0	0.0	16.708
SDQMSC	0.016	0.891	0.0	0.0
APIMSC	3.490	5.418	1.412	0.0
SCAMSC	3.928	3.696	1.412	0.0

PHI

	KSI 1	KSI 2	KSI 3	KSI 4
KSI 1	0.0			
KSI 2	0.0	0.0		
KSI 3	0.0	0.0	0.0	
KSI 4	0.0	0.0	0.0	0.0

TABLE 3.9. Continued

THETA DELTA

	SDQGSC	APIGSC	SESGSC	SDQASC	SCAASC
SDQGSC	0.0				
APIGSC	0.153	0.0			
SESGSC	0.630	1.188	0.0		
SDQASC	0.0	2.873	0.813	0.0	
SCAASC	1.080	5.631	0.380	8.793	0.0
SDQESC	2.548	1.500	0.258	2.777	0.635
APIESC	0.181	0.0	0.322	0.152	2.547
SCAESC	3.051	1.194	0.688	15.893	0.0
SDQMSC	2.725	1.239	0.823	0.010	0.340
APIMSC	4.703	0.0	8.764	0.037	2.683
SCAMSC	0.065	4.617	4.097	1.386	0.0

SDQESC	APIESC	SCAESC	SDQMSC	APIMSC
0.0				
15.009	0.0			
2.819	0.606	0.0		
1.166	3.590	0.045	0.0	
1.326	0.0	4.502	3.402	0.0
0.025	0.082	0.0	5.334	0.881

THETA DELTA

	SCAMSC
SCAMSC	0.0

MAXIMUM MODIFICATION INDEX IS   16.71 FOR ELEMENT (8, 4) OF LAMBDA X

STANDARDIZED SOLUTION

LAMBDA X

	KSI 1	KSI 2	KSI 3	KSI 4
SDQGSC	0.928	-0.097	0.0	0.0
APIGSC	0.703	0.0	0.0	0.0
SESGSC	0.886	0.0	0.0	0.0
SDQASC	0.0	0.836	0.0	0.0
SCAASC	0.0	0.812	0.0	0.0
SDQESC	0.180	0.0	0.735	0.0
APIESC	0.0	0.0	0.892	0.0
SCAESC	0.0	0.182	0.623	0.0
SDQMSC	0.0	0.0	-0.045	0.942
APIMSC	0.0	0.0	0.0	0.908
SCAMSC	0.0	0.0	0.0	0.872

PHI

	KSI 1	KSI 2	KSI 3	KSI 4
KSI 1	1.000			
KSI 2	0.415	1.000		
KSI 3	0.222	0.542	1.000	
KSI 4	0.279	0.632	0.035	1.000

THE PROBLEM REQUIRED 1915 DOUBLE PRECISION WORDS,
THE CPU-TIME WAS   1.82 SECONDS

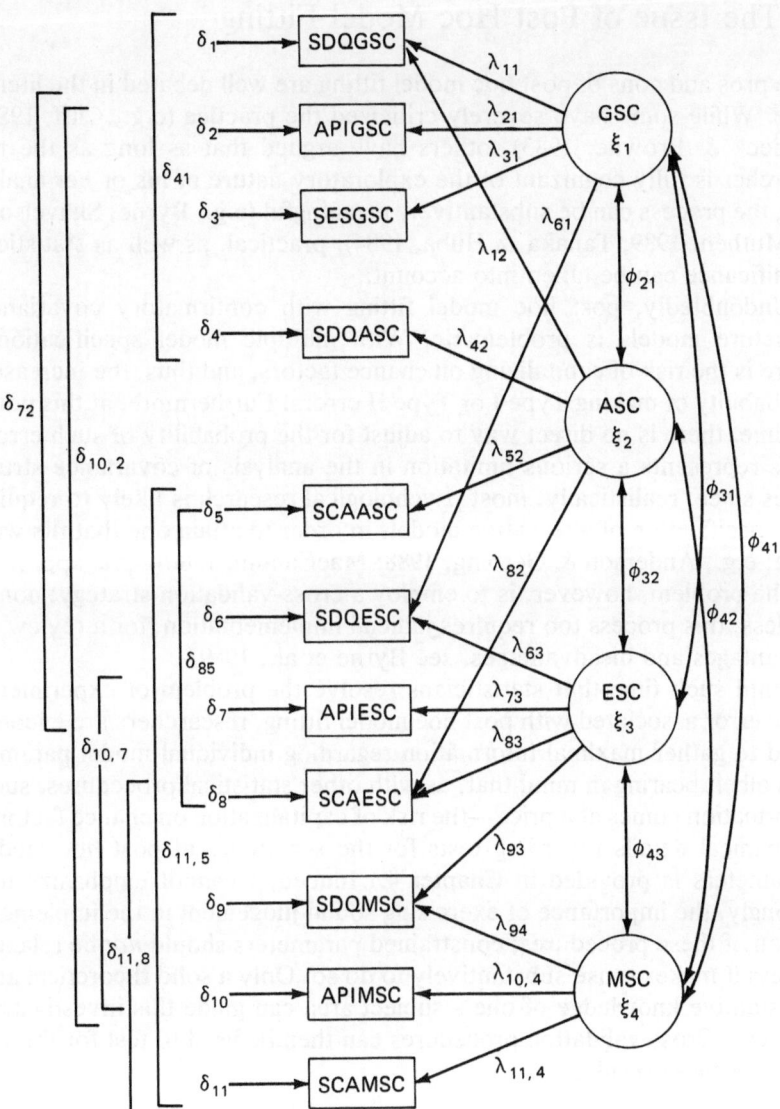

FIGURE 3.2. Structure of Final Four-Factor Model of Self-Concept.

3. The Issue of Post Hoc Model Fitting

The pros and cons of post hoc model fitting are well debated in the literature. While some have severely criticized the practice (e.g., Cliff, 1983; Cudeck & Browne, 1983), others have argued that as long as the researcher is fully cognizant of the exploratory nature of his or her analyses, the process can be substantively meaningful (e.g., Byrne, Shavelson, & Muthén, 1989; Tanaka & Huba, 1984); practical, as well as statistical significance can be taken into account.

Undoubtedly, post hoc model fitting with confirmatory covariance structure models is problematic. With multiple model specifications, there is the risk of capitalizing on chance factors, and thus, the increased probability of making Type I or Type II errors. Furthermore, at this point in time, there is no direct way to adjust for the probability of such error. This represents a serious limitation in the analysis of covariance structures since, realistically, most psychological research is likely to require the specification of alternative models in order to attain one that fits well (see, e.g., Anderson & Gerbing, 1988; MacCallum, 1986). One approach to the problem, however, is to employ a cross-validation strategy; nonetheless, this process too requires judicial implementation (for a review of advantages and disadvantages, see Byrne et al., 1989).

Until such time that statisticians resolve the problem of experiment-wise error associated with post hoc model fitting, researchers are encouraged to gather maximal information regarding individual model parameters albeit bearing in mind that, as with other statistical procedures, such information comes at a price—the risk of capitalization on chance factors. (Technical details regarding tests for the sensitivity of post hoc model parameters is provided in Chapter 4.) Indeed, I cannot emphasize too strongly, the importance of exercising sound judgement in the implementation of these procedures; constrained parameters should *not* be relaxed unless it makes sense substantively to do so. Only a solid theoretical and substantive knowledge of one's subject area can guide this investigative process. Cross-validation procedures can then be used to test for the validity of these results.

Hypothesis 2: Self-Concept Is a Two-Factor Structure

The model to be tested in this hypothesis postulates a priori that SC is a two-factor structure consisting of GSC and ASC. As such, all three GSC measures load onto the GSC factor, while all other measures load onto the ASC factor. This model argues against the viability of subject-specific SC academic factors. This model is schematically represented in Figure 3.3.

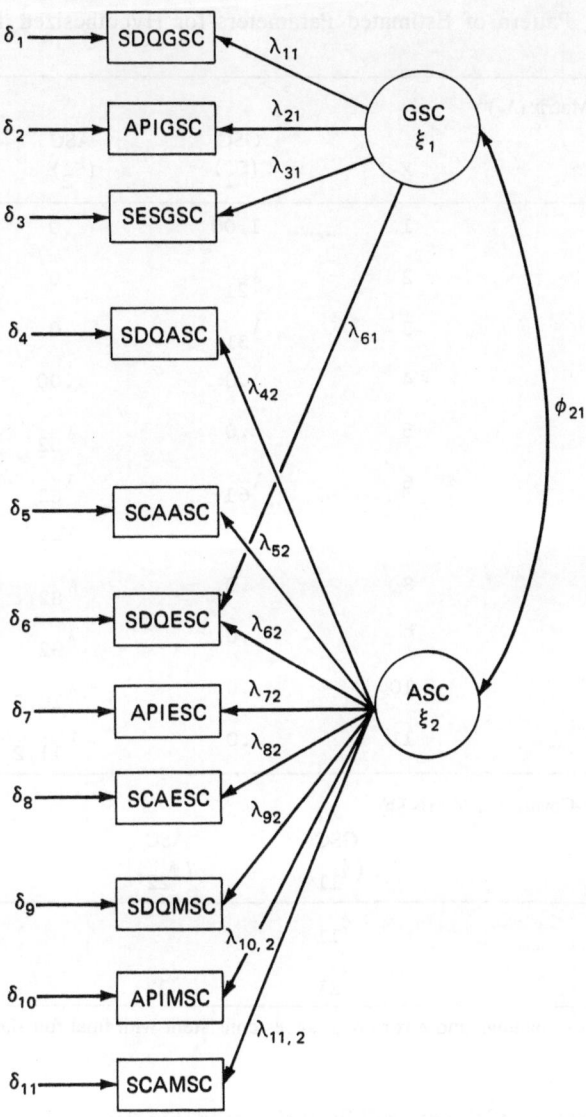

FIGURE 3.3. Hypothesized Structure of Two-Factor Model of Self-Concept.

As with our four-factor model, we will again examine the pattern of specified factor loadings, variance-covariances, and error variances. For purposes of comparison, relevant cross-loadings and all error covariances specified in the four-factor model were similarly specified in the two-factor model. This specification is presented in Table 3.10 and the accompanying LISREL input is shown in Table 3.11.

TABLE 3.10. Pattern of Estimated Parameters for Hypothesized Two-Factor Model

Factor Loading Matrix (Λ_x)[a]

SC measure	X	GSC (ξ_1)	ASC (ξ_2)
SDQGSC	1	1.00	.0
APIGSC	2	λ_{21}	.0
SESGSC	3	λ_{31}	.0
SDQASC	4	.0	1.00
SCAASC	5	.0	λ_{52}
SDQESC	6	λ_{61}	λ_{62}
APIESC	7	.0	λ_{72}
SCAESC	8	.0	λ_{82}
SDQMSC	9	.0	λ_{92}
APIMSC	10	.0	$\lambda_{10,2}$
SCAMSC	11	.0	$\lambda_{11,2}$

Factor Variance-Covariance Matrix (Φ)

	GSC (ϕ_{11})	ASC (ϕ_{22})
GSC	ϕ_{11}	
ASC	ϕ_{21}	ϕ_{22}

[a]Secondary factor loadings and error covariances consistent with final four-factor model.

An examination of the overall fit of this model indicates that it is clearly not a good fit to the data. Selected portions of the LISREL output showing the estimates, overall model fit, standard errors, Q-plot of normalized residuals, and MIs are presented in Table 3.12.

TABLE 3.10. Continued

Error Variance-Covariance Matrix (Θ_δ)

	X_1	X_2	X_3	X_4	X_5	X_6	X_7	X_8	X_9	X_{10}	X_{11}
X_1	$\theta_{\delta_{11}}$										
X_2	0	$\theta_{\delta_{22}}$									
X_3	0	0	$\theta_{\delta_{33}}$								
X_4	$\theta_{\delta_{41}}$	0	0	$\theta_{\delta_{44}}$							
X_5	0	0	0	0	$\theta_{\delta_{55}}$						
X_6	0	0	0	0	0	$\theta_{\delta_{66}}$					
X_7	0	$\theta_{\delta_{72}}$	0	0	0	0	$\theta_{\delta_{77}}$				
X_8	0	0	0	0	$\theta_{\delta_{85}}$	0	0	$\theta_{\delta_{88}}$			
X_9	0	0	0	0	0	0	0	0	$\theta_{\delta_{99}}$		
X_{10}	0	$\theta_{\delta_{10,2}}$	0	0	0	0	$\theta_{\delta_{10,7}}$	0	0	$\theta_{\delta_{10,10}}$	
X_{11}	0	0	0	0	$\theta_{\delta_{11,5}}$	0	0	$\theta_{\delta_{11,8}}$	0	0	$\theta_{\delta_{11,11}}$

TABLE 3.11. LISREL Input for Two-Factor Model

```
MO NX=11 NK=2 LX=FU PH=SY TD=SY,FI
FR LX(2,1) LX(3,1) LX(5,2) LX(6,2) LX(7,2) LX(8,2) LX(9,2) LX(10,2)
FR LX(11,2)
FR LX(6,1) LX(8,2) LX(1,2)
FR TD(1,1) TD(2,2) TD(3,3) TD(4,4) TD(5,5) TD(6,6) TD(7,7)
FR TD(8,8) TD(9,9) TD(10,10) TD(11,11)
FR TD(3,5) TD(11,5) TD(10,7) TD(11,8) TD(10,2) TD(4,1) TD(7,2)
ST 1.0 LX(1,1) LX(4,2)
ST .7 LX(2,1) LX(3,1) LX(5,2) LX(6,2) LX(7,2) LX(8,2) LX(9,2)
ST .7 LX(10,2) LX(11,2)
ST .2 LX(6,1) LX(8,2) LX(1,2)
ST .5 PH(1,1) PH(2,2)
ST .2 PH(2,1)
ST .3 TD(1,1) TD(2,2) TD(3,3) TD(4,4) TD(5,5) TD(6,6) TD(7,7)
ST .3 TD(8,8) TD(9,9) TD(10,10) TD(11,11)
ST .1 TD(8,5) TD(11,5) TD(10,7) TD(11,8) TD(10,2) TD(4,1) TD(7,2)
OU NS SE TV RS MI
```

TABLE 3.12. Selected LISREL Output for Two-Factor Model

LISREL ESTIMATES (MAXIMUM LIKELIHOOD)

LAMBDA X

	KSI 1	KSI 2
SDQGSC	1.000	-0.159
APIGSC	0.787	0.0
SESGSC	0.995	0.0
SDQASC	0.0	1.000
SCAASC	0.0	1.086
SDQESC	0.376	-0.166
APIESC	0.0	0.064
SCAESC	0.0	0.160
SDQMSC	0.0	1.974
APIMSC	0.0	1.923
SCAMSC	0.0	1.838

PHI

	KSI 1	KSI 2
KSI 1	0.798	
KSI 2	0.122	0.228

THETA DELTA

	SDQGSC	APIGSC	SESGSC	SDQASC	SCAASC
SDQGSC	0.215				
APIGSC	0.0	0.491			
SESGSC	0.0	0.0	0.210		
SDQASC	0.045	0.0	0.0	0.772	
SCAASC	0.0	0.0	0.0	0.0	0.732
SDQESC	0.0	0.0	0.0	0.0	0.0
APIESC	0.0	0.057	0.0	0.0	0.0
SCAESC	0.0	0.0	0.0	0.0	0.550
SDQMSC	0.0	0.0	0.0	0.0	0.0
APIMSC	0.0	0.053	0.0	0.0	0.0
SCAMSC	0.0	0.0	0.0	0.0	0.123

	SDQESC	APIESC	SCAESC	SDQMSC	APIMSC
SDQESC	0.896				
APIESC	0.0	0.999			
SCAESC	0.0	0.0	0.994		
SDQMSC	0.0	0.0	0.0	0.113	
APIMSC	0.0	0.112	0.0	0.0	0.154
SCAMSC	0.0	0.0	0.079	0.0	0.0

THETA DELTA

	SCAMSC
SCAMSC	0.231

SQUARED MULTIPLE CORRELATIONS FOR X - VARIABLES

SDQGSC	APIGSC	SESGSC	SDQASC	SCAASC
0.785	0.509	0.790	0.228	0.268

SDQESC	APIESC	SCAESC	SDQMSC	APIMSC
0.104	0.001	0.006	0.887	0.846

SQUARED MULTIPLE CORRELATIONS FOR X - VARIABLES

SCAMSC
0.769

TOTAL COEFFICIENT OF DETERMINATION FOR X - VARIABLES
IS 0.960

TABLE 3.12. Continued

```
        MEASURES OF GOODNESS OF FIT FOR THE WHOLE MODEL :

CHI-SQUARE WITH   34 DEGREES OF FREEDOM IS     1895.11
(PROB. LEVEL = 0.0  )

                        GOODNESS OF FIT INDEX IS 0.723

                ADJUSTED GOODNESS OF FIT INDEX IS 0.463

                ROOT MEAN SQUARE RESIDUAL IS        0.190
```

STANDARD ERRORS

LAMBDA X

	KSI 1	KSI 2
SDQGSC	0.0	0.047
APIGSC	0.032	0.0
SESGSC	0.033	0.0
SDQASC	0.0	0.0
SCAASC	0.0	0.088
SDQESC	0.038	0.069
APIESC	0.0	0.069
SCAESC	0.0	0.069
SDQMSC	0.0	0.121
APIMSC	0.0	0.119
SCAMSC	0.0	0.115

PHI

	KSI 1	KSI 2
KSI 1	0.050	
KSI 2	0.018	0.029

THETA DELTA

	SDQGSC	APIGSC	SESGSC	SDQASC	SCAASC
SDQGSC	0.021				
APIGSC	0.0	0.025			
SESGSC	0.0	0.0	0.021		
SDQASC	0.017	0.0	0.0	0.035	
SCAASC	0.0	0.0	0.0	0.0	0.034
SDQESC	0.0	0.0	0.0	0.0	0.0
APIESC	0.0	0.024	0.0	0.0	0.0
SCAESC	0.0	0.0	0.0	0.0	0.033
SDQMSC	0.0	0.0	0.0	0.0	0.0
APIMSC	0.0	0.011	0.0	0.0	0.0
SCAMSC	0.0	0.0	0.0	0.0	0.016

	SDQESC	APIESC	SCAESC	SDQMSC	APIMSC
	0.041				
	0.0	0.045			
	0.0	0.0	0.045		
	0.0	0.0	0.0	0.010	
	0.0	0.016	0.0	0.0	0.011
	0.0	0.0	0.017	0.0	0.0

THETA DELTA

	SCAMSC
SCAMSC	0.013

TABLE 3.12. Continued

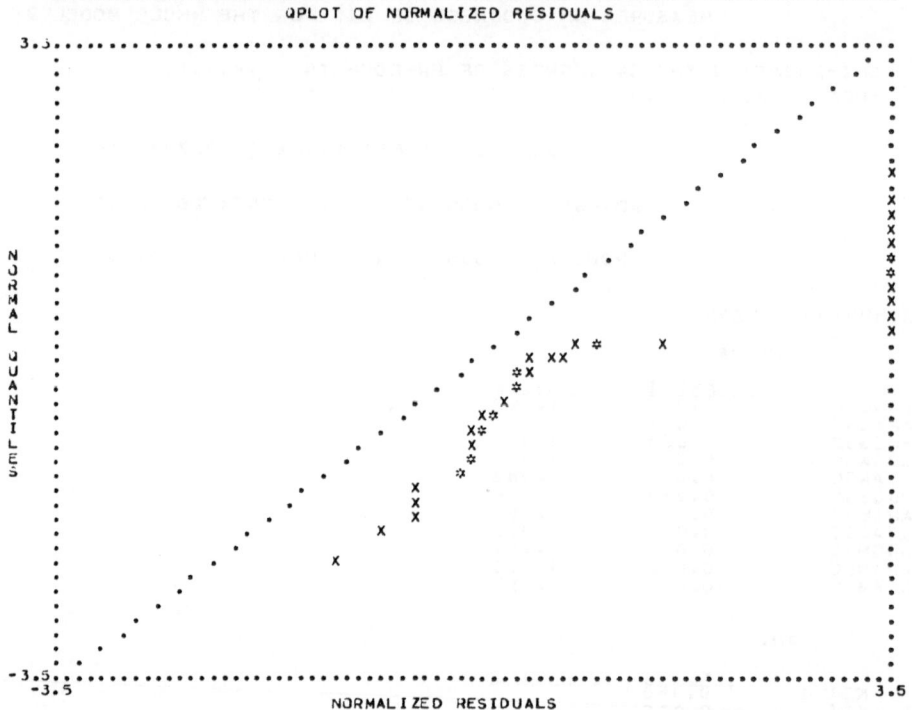

QPLOT OF NORMALIZED RESIDUALS

THE PROBLEM REQUIRED 1501 DOUBLE PRECISION WORDS, THE CPU-TIME WAS   1.63 SECONDS

MODIFICATION INDICES

LAMBDA X

	KSI 1	KSI 2
SDQGSC	0.0	0.0
APIGSC	0.0	1.150
SESGSC	0.0	1.150
SDQASC	63.212	0.0
SCAASC	15.102	0.0
SDQESC	0.0	0.0
APIESC	37.264	0.0
SCAESC	2.686	0.0
SDQMSC	0.416	0.0
APIMSC	0.024	0.0
SCAMSC	14.334	0.0

PHI

	KSI 1	KSI 2
KSI 1	0.0	
KSI 2	0.0	0.0

TABLE 3.12. Continued

THETA DELTA

	SDQGSC	APIGSC	SESGSC	SDQASC	SCAASC
SDQGSC	0.0				
APIGSC	1.782	0.0			
SESGSC	0.098	1.479	0.0		
SDQASC	0.0	0.112	5.890	0.0	
SCAASC	0.226	0.199	9.032	129.836	0.0
SDQESC	1.479	1.437	4.159	150.549	1.380
APIESC	7.202	0.0	2.259	241.917	1.814
SCAESC	4.659	0.008	0.200	28.509	0.0
SDQMSC	2.611	0.606	2.370	0.158	11.380
APIMSC	2.758	0.0	6.123	1.766	0.264
SCAMSC	0.074	2.192	10.336	30.949	0.0

	SDQESC	APIESC	SCAESC	SDQMSC	APIMSC
	0.0				
	473.929	0.0			
	148.642	231.085	0.0		
	0.110	7.072	3.089	0.0	
	8.343	0.0	11.632	0.000	0.0
	3.181	18.510	0.0	8.623	1.015

THETA DELTA

	SCAMSC
SCAMSC	0.0

MAXIMUM MODIFICATION INDEX IS   473.93 FOR ELEMENT (7, 6)
OF THETA DELTA

Hypothesis 3: Self-Concept Is a One-Factor Structure

A review of the SC literature reveals that there are many who still argue
for the unidimensionality of the construct. Thus, it was considered important to test the fit of a one-factor model of SC. This model is presented
schematically in Figure 3.4, the specification of parameters summarized
in Table 3.13, and the LISREL input presented in Table 3.14.

Finally, the selected LISREL output, as with the two-factor model, is
presented in Table 3.15.

Examination of goodness-of-fit indices for both the hypothesized two-
factor and one-factor models of SC reveals a clear indication of misspecified models. Based on these findings, we concluded that SC is a multidimensional construct, which in this study comprised the four facets of
GSC, ASC, ESC, and MSC.

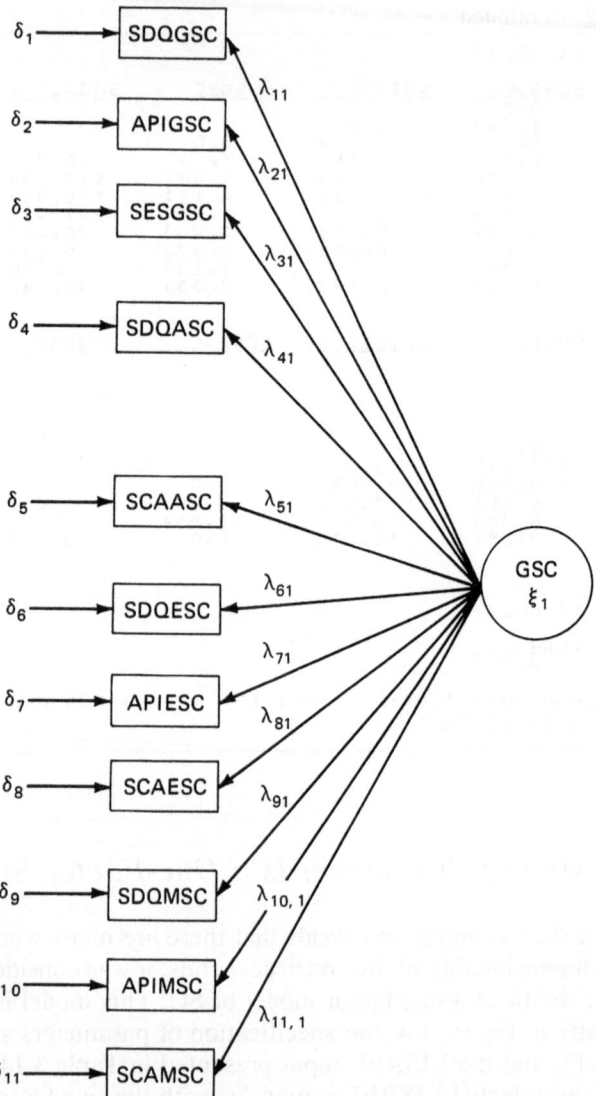

FIGURE 3.4. Hypothesized Structure of One-Factor Model of Self-Concept.

TABLE 3.13. Pattern of Estimated Parameters for Hypothesized One-Factor Model

Factor Loading Matrix (Λ_x)[a]

SC measure	X	GSC (ξ_1)
SDQGSC	X_1	1.00
APIASC	X_2	λ_{21}
SESGSC	X_3	λ_{31}
SDQASC	X_4	λ_{41}
SCAASC	X_5	λ_{51}
SDQESC	X_6	λ_{61}
APIESC	X_7	λ_{71}
SCAESC	X_8	λ_{81}
SDQMSC	X_9	λ_{91}
APIMSC	X_{10}	$\lambda_{10,1}$
SCAMSC	X_{11}	$\lambda_{11,1}$

Factor Variance-Covariance Matrix (Φ)

	GSC (ϕ_{11})
GSC	ϕ_{11}

[a]Secondary factor loadings and error covariances consistent with final four-factor model.

TABLE 3.13. Continued

Error Variance-Covariance Matrix (Θ_δ)

	X_1	X_2	X_3	X_4	X_5	X_6	X_7	X_8	X_9	X_{10}	X_{11}
X_1	$\theta_{\delta_{11}}$										
X_2	0	$\theta_{\delta_{22}}$									
X_3	0	0	$\theta_{\delta_{33}}$								
X_4	$\theta_{\delta_{41}}$	0	0	$\theta_{\delta_{44}}$							
X_5	0	0	0	0	$\theta_{\delta_{55}}$						
X_6	0	0	0	0	0	$\theta_{\delta_{66}}$					
X_7	0	$\theta_{\delta_{72}}$	0	0	0	0	$\theta_{\delta_{77}}$				
X_8	0	0	0	0	$\theta_{\delta_{85}}$	0	0	$\theta_{\delta_{88}}$			
X_9	0	0	0	0	0	0	0	0	$\theta_{\delta_{99}}$		
X_{10}	0	$\theta_{\delta_{10,2}}$	0	0	0	0	$\theta_{\delta_{10,7}}$	0	0	$\theta_{\delta_{10,10}}$	
X_{11}	0	0	0	0	$\theta_{\delta_{11,5}}$	0	0	$\theta_{\delta_{11,8}}$	0	0	$\theta_{\delta_{11,11}}$

TABLE 3.14. LISREL Input for One-Factor Model

```
MO NX=11 NK=1 LX=FU PH=SY TD=SY,FI
FR LX(2,1) LX(3,1) LX(4,1) LX(5,1) LX(6,1) LX(7,1) LX(8,1) LX(9,1)
FR LX(10,1) LX(11,1)
FR LX(6,1)
FR TD(1,1) TD(2,2) TD(3,3) TD(4,4) TD(5,5) TD(6,6) TD(7,7)
FR TD(8,8) TD(9,9) TD(10,10) TD(11,11)
FR TD(8,5) TD(11,5) TD(10,7) TD(11,8) TD(10,2) TD(4,1) TD(7,2)
ST 1.0 LX(1,1)
ST .7 LX(2,1) LX(3,1) LX(4,1) LX(5,1) LX(6,1) LX(7,1) LX(8,1) LX(9,1)
ST .7 LX(10,1) LX(11,1)
ST .2 LX(6,1)
ST .5 PH(1,1)
ST .3 TD(1,1) TD(2,2) TD(3,3) TD(4,4) TD(5,5) TD(6,6) TD(7,7)
ST .3 TD(8,8) TD(9,9) TD(10,10) TD(11,11)
ST .1 TD(8,5) TD(11,5) TD(10,7) TD(11,8) TD(10,2) TD(4,1) TD(7,2)
OU NS SE TV RS MI
```

TABLE 3.15. Selected LISREL Output for One-Factor Model

LISREL ESTIMATES (MAXIMUM LIKELIHOOD)

      LAMBDA X

	KSI 1
SDQGSC	1.000
APIGSC	1.273
SESGSC	1.346
SDQASC	2.571
SCAASC	2.785
SDQESC	0.127
APIESC	0.208
SCAESC	0.433
SDQMSC	5.031
APIMSC	4.919
SCAMSC	4.692

      PHI

	KSI 1
KSI 1	0.035

THETA DELTA

	SDQGSC	APIGSC	SESGSC	SDQASC	SCAASC
SDQGSC	0.965				
APIGSC	0.0	0.943			
SESGSC	0.0	0.0	0.937		
SDQASC	0.211	0.0	0.0	0.769	
SCAASC	0.0	0.0	0.0	0.0	0.729
SDQESC	0.0	0.0	0.0	0.0	0.0
APIESC	0.0	0.181	0.0	0.0	0.0
SCAESC	0.0	0.0	0.0	0.0	0.547
SDQMSC	0.0	0.0	0.0	0.0	0.0
APIMSC	0.0	0.051	0.0	0.0	0.0
SCAMSC	0.0	0.0	0.0	0.0	0.120

	SDQESC	APIESC	SCAESC	SDQMSC	APIMSC
SDQESC	0.999				
APIESC	0.0	0.998			
SCAESC	0.0	0.0	0.993		
SDQMSC	0.0	0.0	0.0	0.114	
APIMSC	0.0	0.106	0.0	0.0	0.153
SCAMSC	0.0	0.0	0.075	0.0	0.0

THETA DELTA

	SCAMSC
SCAMSC	0.230

SQUARED MULTIPLE CORRELATIONS FOR X - VARIABLES

SDQGSC	APIGSC	SESGSC	SDQASC	SCAASC
0.035	0.057	0.063	0.231	0.271

SDQESC	APIESC	SCAESC	SDQMSC	APIMSC
0.001	0.002	0.007	0.886	0.847

SQUARED MULTIPLE CORRELATIONS FOR X - VARIABLES

SCAMSC
0.770

TOTAL COEFFICIENT OF DETERMINATION FOR X - VARIABLES
IS -0.635

TABLE 3.15. Continued

MEASURES OF GOODNESS OF FIT FOR THE WHOLE MODEL :

CHI-SQUARE WITH 37 DEGREES OF FREEDOM IS   3392.96
(PROB. LEVEL = 0.0)

GOODNESS OF FIT INDEX IS 0.605

ADJUSTED GOODNESS OF FIT INDEX IS 0.295

ROOT MEAN SQUARE RESIDUAL IS   0.238

STANDARD ERRORS

LAMBDA X

	KSI 1
SDQGSC	0.0
APIGSC	0.277
SESGSC	0.286
SDQASC	0.429
SCAASC	0.500
SDQESC	0.176
APIESC	0.180
SCAESC	0.190
SDQMSC	0.864
APIMSC	0.845
SCAMSC	0.808

PHI

	KSI 1
KSI 1	0.012

THETA DELTA

	SDQGSC	APIGSC	SESGSC	SDQASC	SCAASC
SDQGSC	0.043				
APIGSC	0.0	0.043			
SESGSC	0.0	0.0	0.042		
SDQASC	0.028	0.0	0.0	0.035	
SCAASC	0.0	0.0	0.0	0.0	0.034
SDQESC	0.0	0.0	0.0	0.0	0.0
APIESC	0.0	0.031	0.0	0.0	0.0
SCAESC	0.0	0.0	0.0	0.0	0.032
SDQMSC	0.0	0.0	0.0	0.0	0.0
APIMSC	0.0	0.015	0.0	0.0	0.0
SCAMSC	0.0	0.0	0.0	0.0	0.015

	SDQESC	APIESC	SCAESC	SDQMSC	APIMSC
SDQESC	0.045				
APIESC	0.0	0.045			
SCAESC	0.0	0.0	0.045		
SDQMSC	0.0	0.0	0.0	0.010	
APIMSC	0.0	0.015	0.0	0.0	0.011
SCAMSC	0.0	0.0	0.017	0.0	0.0

THETA DELTA

	SCAMSC
SCAMSC	0.013

TABLE 3.15. Continued

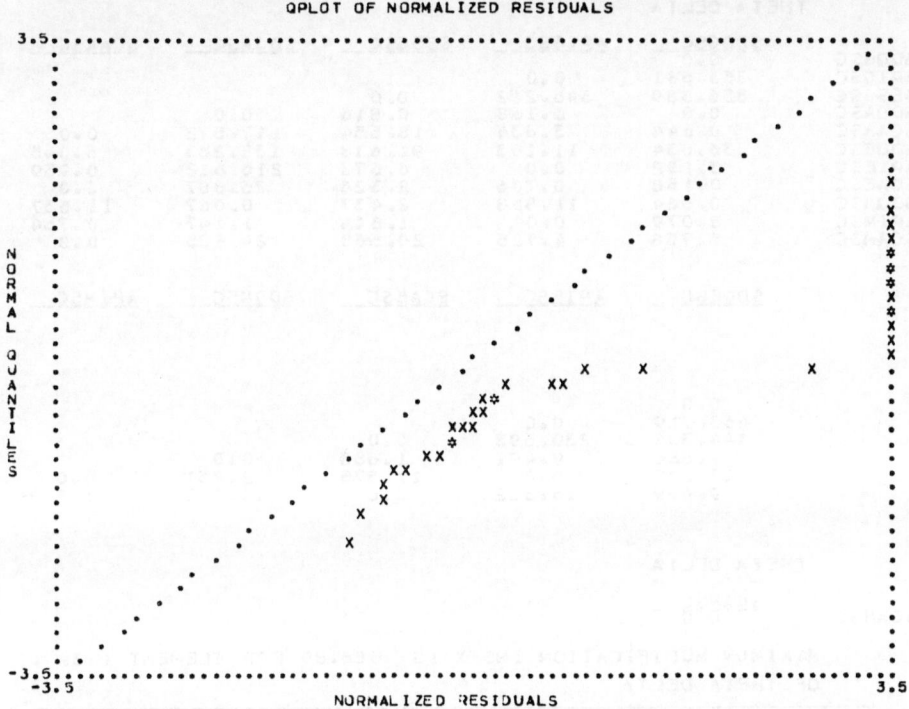

THE PROBLEM REQUIRED 1336 DOUBLE PRECISION WORDS, THE CPU-TIME WAS   1.54 SECONDS

MODIFICATION INDICES

LAMBDA X

	KSI 1
SDQGSC	0.0
APIGSC	0.0
SESGSC	0.0
SDQASC	0.0
SCAASC	0.0
SDQESC	0.0
APIESC	0.0
SCAESC	0.0
SDQMSC	0.0
APIMSC	0.0
SCAMSC	0.0

PHI

	KSI 1
KSI 1	0.0

TABLE 3.15. Continued

THETA DELTA

	SDQGSC	APIGSC	SESGSC	SDQASC	SCAASC
SDQGSC	0.0				
APIGSC	353.681	0.0			
SESGSC	556.889	346.212	0.0		
SDQASC	0.0	3.168	0.816	0.0	
SCAASC	0.644	3.834	18.554	117.572	0.0
SDQESC	36.034	11.183	91.618	138.263	5.045
APIESC	3.152	0.0	6.573	215.612	0.959
SCAESC	0.158	0.706	2.326	25.887	0.0
SDQMSC	0.664	11.868	2.437	0.067	11.637
APIMSC	8.079	0.0	1.844	1.147	0.764
SCAMSC	5.758	4.725	20.569	24.185	0.0

	SDQESC	APIESC	SCAESC	SDQMSC	APIMSC
	0.0				
	463.719	0.0			
	144.304	230.598	0.0		
	1.686	9.491	1.683	0.0	
	12.300	0.0	11.375	2.861	0.0
	9.459	16.332	0.0	10.037	3.240

THETA DELTA

	SCAMSC
SCAMSC	0.0

MAXIMUM MODIFICATION INDEX IS 556.89 FOR ELEMENT (3, 1) OF THETA DELTA

Summary

This chapter presented an application of LISREL CFA in testing for the factorial structure of a theoretical construct. Specifically, a four-factor model of SC was tested against competing two- and one-factor models. For each model, the LISREL input was presented, together with related tabular and schematic illustrations of relations among the variables. A thorough review of the LISREL output provided a guide to the interpretation of model fit with respect to the model as a whole, and for individual model parameters. Finally, problems associated with post hoc model fitting were addressed, and caveats issued regarding the importance of selecting substantively meaningful final models.

4
Application 2: Validating a Measuring Instrument

Our second application tests hypotheses related to the Self Description Questionnaire III (SDQIII; Marsh & O'Neill, 1984), an instrument designed to measure 13 facets of SC: one general SC, three academic SCs (English, mathematics, and general school), and nine nonacademic SCs (physical ability, physical appearance, social (same sex), social (opposite sex), parent relations, emotional stability, problem solving/creative thinking, religion/spirituality, and honesty/reliability). Of relevance to the present application is the factorial validity of the general and academic SC subscales for males only; tests of hypotheses related to factorial invariance will be addressed in Section III. (For details of the study related to this application, see Byrne, 1988b.)

1. The SDQIII: The Measuring Instrument Under Study

The SDQIII is composed of 136 items that are structured on an eight-point likert-type scale with responses ranging from "1—Definitely False" to "8—Definitely True." The General-self subscale contains 12 items and was used to measure general SC. The Academic SC, Verbal SC, and Mathematics SC subscales each contain 10 items and were used to measure general school, English, and mathematics SCs, respectively.

In testing for the factorial validity of a measuring instrument using CFA procedures, the researcher seeks to determine the extent to which items designed to assess a particular factor (i.e., facet or dimension of a construct) actually do so. In general, subscales of a measuring instrument are considered to represent the factors of a construct; all items in a particular subscale are therefore expected to load onto that factor.

In the present application, all analyses were based on item pairs, rather than on single items. (An elaboration of the rationale underlying this procedure, as well as the method of item-pair formation, is detailed in the reference article cited for this chapter.) The CFA model hypothesized a priori that: responses to the SDQIII could be explained by four factors

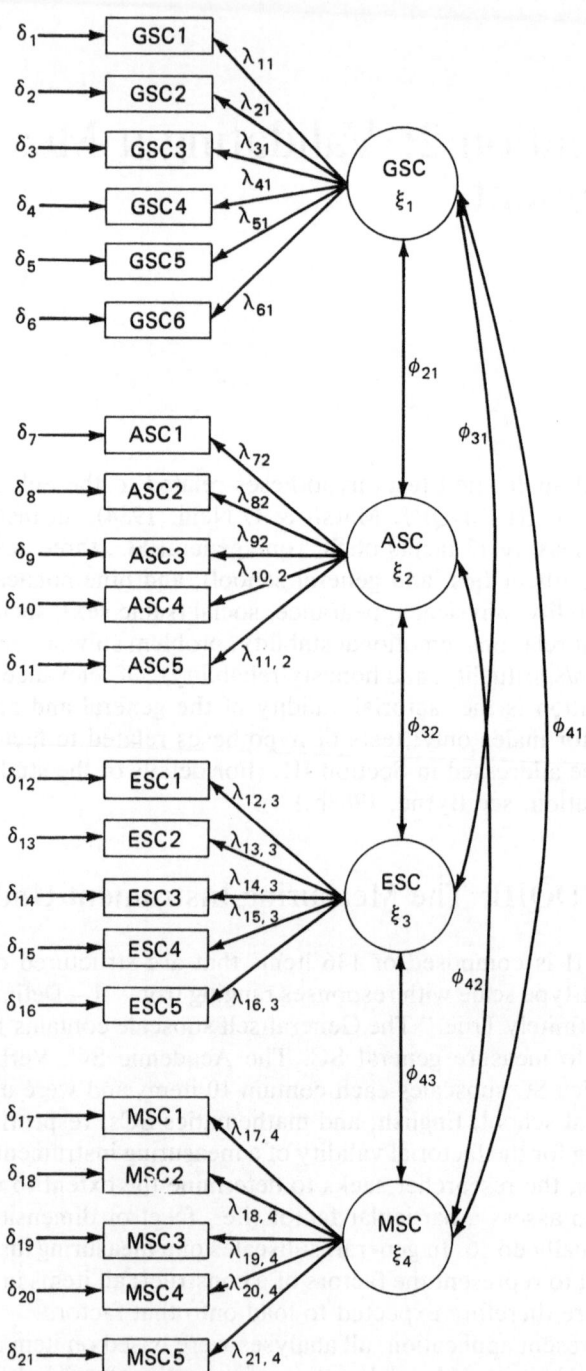

FIGURE 4.1. Hypothesized Four-Factor Model of Self-Concept for the SDQIII.

TABLE 4.1. Pattern of Estimated Parameters for Hypothesized Four-Factor Model for the SDQIII

Factor Loading Matrix (Λ_x)

Measure	X	GSC (ξ_1)	ASC (ξ_2)	ESC (ξ_3)	MSC (ξ_4)
GSC1	1	1.00	.0	.0	.0
GSC2	2	λ_{21}	.0	.0	.0
GSC3	3	λ_{31}	.0	.0	.0
GSC4	4	λ_{41}	.0	.0	.0
GSC5	5	λ_{51}	.0	.0	.0
GSC6	6	λ_{61}	.0	.0	.0
ASC1	7	.0	1.00	.0	.0
ASC2	8	.0	λ_{82}	.0	.0
ASC3	9	.0	λ_{92}	.0	.0
ASC4	10	.0	$\lambda_{10,2}$	.0	.0
ASC5	11	.0	$\lambda_{11,2}$	.0	.0
ESC1	12	.0	.0	1.00	.0
ESC2	13	.0	.0	$\lambda_{13,3}$	.0
ESC3	14	.0	.0	$\lambda_{14,3}$	.0
ESC4	15	.0	.0	$\lambda_{15,3}$	.0
ESC5[a]	16	.0	.0	$\lambda_{16,3}$	.0
MSC1	17	.0	.0	.0	1.00
MSC2	18	.0	.0	.0	$\lambda_{18,4}$
MSC3	19	.0	.0	.0	$\lambda_{19,4}$
MSC4	20	.0	.0	.0	$\lambda_{20,4}$
MSC5	21	.0	.0	.0	$\lambda_{21,4}$

[a]ESC5 was composed of item 9 only on the Verbal SC subscale.
GSC1 to GSC6 = item-pairs 1 & 2, 3 & 4, 5 & 6, 7 & 8, 9 & 10, and 11 & 12 of the General-Self subscale; ASC1 to ASC5 = item-pairs 1 & 2, 3 & 4, 5 & 6, 7 & 8, and 9 & 10 of the Academic SC subscale; ESC1 to ESC4 = item-pairs 1 & 2, 3 & 4, 5 & 6, 7 & 8 of the Verbal SC subscale; MSC1 to MSC5 = item-pairs 1 & 2, 3 & 4, 5 & 6, 7 & 8, and 9 & 10 of the Mathematics SC subscale.

TABLE 4.1. Continued

Factor Variance-Covariance Matrix (Φ)

	GSC (ϕ_{11})	ASC (ϕ_{22})	ESC (ϕ_{33})	MSC (ϕ_{44})
GSC	ϕ_{11}			
ASC	ϕ_{21}	ϕ_{22}		
ESC	ϕ_{31}	ϕ_{32}	ϕ_{33}	
MSC	ϕ_{41}	ϕ_{42}	ϕ_{43}	ϕ_{44}

Error Variance-Covariance Matrix (Θ_δ)

	x_1	x_2	x_3	x_4	x_5	x_6	x_7	x_8	x_9	x_{10}	x_{11}	x_{12}	x_{13}
x_1	θ_{δ_1}	.0	.0	.0	.0	.0	.0	.0	.0	.0	.0	.0	.0
x_2	.0	θ_{δ_2}	.0	.0	.0	.0	.0	.0	.0	.0	.0	.0	.0
x_3	.0	.0	θ_{δ_3}	.0	.0	.0	.0	.0	.0	.0	.0	.0	.0
x_4	.0	.0	.0	θ_{δ_4}	.0	.0	.0	.0	.0	.0	.0	.0	.0
x_5	.0	.0	.0	.0	θ_{δ_5}	.0	.0	.0	.0	.0	.0	.0	.0
x_6	.0	.0	.0	.0	.0	θ_{δ_6}	.0	.0	.0	.0	.0	.0	.0
x_7	.0	.0	.0	.0	.0	.0	θ_{δ_7}	.0	.0	.0	.0	.0	.0
x_8	.0	.0	.0	.0	.0	.0	.0	θ_{δ_8}	.0	.0	.0	.0	.0
x_9	.0	.0	.0	.0	.0	.0	.0	.0	θ_{δ_9}	.0	.0	.0	.0
x_{10}	.0	.0	.0	.0	.0	.0	.0	.0	.0	$\theta_{\delta_{10}}$	.0	.0	.0
x_{11}	.0	.0	.0	.0	.0	.0	.0	.0	.0	.0	$\theta_{\delta_{11}}$	.0	.0
x_{12}	.0	.0	.0	.0	.0	.0	.0	.0	.0	.0	.0	$\theta_{\delta_{12}}$	.0
x_{13}	.0	.0	.0	.0	.0	.0	.0	.0	.0	.0	.0	.0	$\theta_{\delta_{13}}$
x_{14}	.0	.0	.0	.0	.0	.0	.0	.0	.0	.0	.0	.0	.0
x_{15}	.0	.0	.0	.0	.0	.0	.0	.0	.0	.0	.0	.0	.0
x_{16}	.0	.0	.0	.0	.0	.0	.0	.0	.0	.0	.0	.0	.0
x_{17}	.0	.0	.0	.0	.0	.0	.0	.0	.0	.0	.0	.0	.0
x_{18}	.0	.0	.0	.0	.0	.0	.0	.0	.0	.0	.0	.0	.0
x_{19}	.0	.0	.0	.0	.0	.0	.0	.0	.0	.0	.0	.0	.0
x_{20}	.0	.0	.0	.0	.0	.0	.0	.0	.0	.0	.0	.0	.0
x_{21}	.0	.0	.0	.0	.0	.0	.0	.0	.0	.0	.0	.0	.0

(general SC, academic SC, English SC, and mathematics SC), each item-pair would have a nonzero loading on the SC factor it was designed to measure (i.e., the target factor), and zero loadings on all other factors (i.e., nontarget factors), the four factors would be correlated, and the uniquenesses for the item-pair variables would be uncorrelated. This model is presented schematically in Figure 4.1, and the pattern of parameters to be estimated is detailed in Table 4.1.

In Table 4.1 we see again that, for purposes of statistical identification, the first λ of each set of λs designed to measure the same factor was fixed to 1.0.

X_{14}	X_{15}	X_{16}	X_{17}	X_{18}	X_{19}	X_{20}	X_{21}
.0	.0	.0	.0	.0	.0	.0	.0
.0	.0	.0	.0	.0	.0	.0	.0
.0	.0	.0	.0	.0	.0	.0	.0
.0	.0	.0	.0	.0	.0	.0	.0
.0	.0	.0	.0	.0	.0	.0	.0
.0	.0	.0	.0	.0	.0	.0	.0
.0	.0	.0	.0	.0	.0	.0	.0
.0	.0	.0	.0	.0	.0	.0	.0
.0	.0	.0	.0	.0	.0	.0	.0
.0	.0	.0	.0	.0	.0	.0	.0
.0	.0	.0	.0	.0	.0	.0	.0
.0	.0	.0	.0	.0	.0	.0	.0
.0	.0	.0	.0	.0	.0	.0	.0
$\theta_{\delta 14}$	.0	.0	.0	.0	.0	.0	.0
.0	$\theta_{\delta 15}$	.0	.0	.0	.0	.0	.0
.0	.0	$\theta_{\delta 16}$	.0	.0	.0	.0	.0
.0	.0	.0	$\theta_{\delta 17}$	.0	.0	.0	.0
.0	.0	.0	.0	$\theta_{\delta 18}$	.0	.0	.0
.0	.0	.0	.0	.0	$\theta_{\delta 19}$	.0	.0
.0	.0	.0	.0	.0	.0	$\theta_{\delta 20}$	.0
.0	.0	.0	.0	.0	.0	.0	$\theta_{\delta 21}$

2. LISREL Input

As in Chapter 3, we'll now transform the model specification into LIS-REL input statements; this information, together with the data presented in correlation matrix form, is presented in Table 4.2.

Let us now review this model specification in light of our LISREL input statements.

2.1. The DA Card

The DA card tells us that there are 21 variables (NI = 21) [six item-pairs measuring general SC and five item-pairs measuring each of the three academic SCs], the sample size is 455 (NO = 455), and the data are to be analyzed as a correlation matrix (MA = KM). The input data are in the form of a symmetric correlation matrix (KM SY).

TABLE 4.2. LISREL Input for Hypothesized Four-Factor Model of Self-Concept for the SDQIII

```
CONFIRMATORY FACTOR ANALYSIS OF SDQ FOR MALES
DA NI=21 NO=455 MA=KM
LA
*
'GSC1' 'GSC2' 'GSC3' 'GSC4' 'GSC5' 'GSC6' 'ASC1' 'ASC2' 'ASC3' 'ASC4'
'ASC5' 'ESC1' 'ESC2' 'ESC3' 'ESC4' 'ESC5' 'MSC1' 'MSC2' 'MSC3' 'MSC4'
'MSC5'
KM SY
(21F3.2)
100U
63100
73 63100
60 51 54100
70 66 73 66100
61 53 58 60 60100
24 20 16 26 17 25100
23 18 17 27 17 19 65100
25 20 22 28 19 22 57 77100
29 24 26 31 25 24 56 68 72100
27 21 21 26 19 27 48 63 73 63100
18 05 13 07 13 13 15 20 22 23 23100
19 18 16 20 21 09 15 25 22 30 20 36100
21 18 26 17 23 24 14 23 21 29 27 38 40100
26 21 29 19 28 22 17 25 25 38 29 41 45 37100
16 09 08 22 15 11 11 18 18 21 18 26 37 20 26100
16 11 12 20 13 12 46 50 45 51 43 04 10 07 14 16100
19 16 19 16 17 13 27 30 36 41-07 03 03 11 07 04100
18 13 16 13 17 14 32 46 49 46 49-02 12 07 13 13 69 77100
20 14 19 16 15 19 34 43 48 47 54 01 10 09 15 09 64 77 79100
15 11 13 14 09 15 37 42 45 46 30 01 08 06 20 08 64 70 67 73100
MO NX=21 NK=4 LX=FU PH=SY TD=SY,FI
FR LX(2,1) LX(3,1) LX(4,1) LX(5,1) LX(6,1) LX(8,2) LX(9,2) LX(10,2)
FR LX(11,2) LX(13,3) LX(14,3) LX(15,3) LX(16,3) LX(18,4) LX(19,4)
FR LX(20,4) LX(21,4)
FR TD(1,1) TD(2,2) TD(3,3) TD(4,4) TD(5,5) TD(6,6) TD(7,7) TD(8,8)
FR TD(9,9) TD(10,10) TD(11,11) TD(12,12) TD(13,13) TD(14,14) TD(15,15)
FR TD(16,16) TD(17,17) TD(18,18) TD(19,19 TD(20,20) TD(21,21)
ST 1.0 LX(1,1) LX(7,2) LX(12,3) LX(17,4)
ST .7 LX(2,1) LX(3,1) LX(4,1) LX(5,1) LX(6,1) LX(8,2) LX(9,2) LX(10,2)
ST .7 LX(11,2) LX(13,3) LX(14,3) LX(15,3) LX(16,3) LX(18,4) LX(19,4)
ST .7 LX(20,4) LX(21,4)
ST .5 PH(1,1) PH(2,2) PH(3,3) PH(4,4)
ST .3 PH(2,1) PH(3,1) PH(3,2) PH(4,1) PH(4,2) PH(4,3)
ST .2 TD(1,1) TD(2,2) TD(3,3) TD(4,4) TD(5,5) TD(6,6) TD(7,7) TD(8,8)
ST .2 TD(9,9) TD(10,10) TD(11,11) TD(12,12) TD(13,13) TD(14,14) TD(15,15)
ST .2 TD(16,16) TD(17,17) TD(18,18) TD(19,19 TD(20,20) TD(21,21)
OU NS SE TV MI SS
```

2.2. The MO Card

The MO card tells us that we are working with an all-X model that consists of 21 X variables (NX = 21) and four latent (ξ) factors (NK = 4), the factor loading matrix is full and fixed (LX = FU), the factor variance/covariance matrix is symmetric and free (PH = SY),[1] and the error variance/covariance matrix is symmetric and fixed (TD = SY,FI). Note that although the TD matrix has been specified as symmetric and fixed, each error variance has been specified as a free parameter [TD(1,1)–TD(21,21)]; no error covariances have been specified, thus making this specification equivalent to a diagonal matrix (TD = DI,FR). Specification of the error matrix as TD = SY,FI, however, acts as a time-saver in the event that the researcher wishes to conduct post hoc analyses that include correlated errors.

2.3. The OU Card

The OU card has specified that no start values are to be provided by the program (NS) and that standard errors (SE), T-values (TV), modification indices (MI), and a standardized solution (SS) are requested in the output.

3. LISREL Output

Discussion of the results will focus on four major aspects of the printout: feasibility of the parameter estimates, adequacy of the measurement model, goodness-of-fit of the overall model, and goodness-of-fit of individual parameter estimates; only portions of the printout related to these phenomena are tabled.

3.1. Feasibility of Parameter Estimates

The parameter estimates and standard estimates for each of the three matrices are presented in Table 4.3.

An examination of Table 4.3 reveals all estimates to be acceptable; all estimates are positive, no correlations are greater than 1.00 and all matrices have positive definite status (i.e., there was no error message indicating that any matrix was not positive definite), and finally, the standard errors range from 0.019 to 0.115.

---

[1]By default, the LX matrix is specified as fixed, and the PH matrix as free.

TABLE 4.3. Selected LISREL Output: Parameter Estimates and Standard Errors

```
LISREL ESTIMATES (MAXIMUM LIKELIHOOD)
```

LAMBDA X

	KSI_1	KSI_2	KSI_3	KSI_4
GSC1	1.000	0.0	0.0	0.0
GSC2	0.889	0.0	0.0	0.0
GSC3	0.991	0.0	0.0	0.0
GSC4	0.861	0.0	0.0	0.0
GSC5	1.024	0.0	0.0	0.0
GSC6	0.855	0.0	0.0	0.0
ASC1	0.0	1.000	0.0	0.0
ASC2	0.0	1.259	0.0	0.0
ASC3	0.0	1.322	0.0	0.0
ASC4	0.0	1.246	0.0	0.0
ASC5	0.0	1.204	0.0	0.0
ESC1	0.0	0.0	1.000	0.0
ESC2	0.0	0.0	1.122	0.0
ESC3	0.0	0.0	0.993	0.0
ESC4	0.0	0.0	1.132	0.0
ESC5	0.0	0.0	0.739	0.0
MSC1	0.0	0.0	0.0	1.000
MSC2	0.0	0.0	0.0	1.124
MSC3	0.0	0.0	0.0	1.159
MSC4	0.0	0.0	0.0	1.169
MSC5	0.0	0.0	0.0	1.057

PHI

	KSI_1	KSI_2	KSI_3	KSI_4
KSI 1	0.708			
KSI 2	0.189	0.447		
KSI 3	0.192	0.184	0.350	
KSI 4	0.147	0.315	0.070	0.581

THETA DELTA

	GSC1	GSC2	GSC3	GSC4	GSC5
GSC1	0.292				
GSC2	0.0	0.441			
GSC3	0.0	0.0	0.305		
GSC4	0.0	0.0	0.0	0.475	
GSC5	0.0	0.0	0.0	0.0	0.257
GSC6	0.0	0.0	0.0	0.0	0.0
ASC1	0.0	0.0	0.0	0.0	0.0
ASC2	0.0	0.0	0.0	0.0	0.0
ASC3	0.0	0.0	0.0	0.0	0.0
ASC4	0.0	0.0	0.0	0.0	0.0
ASC5	0.0	0.0	0.0	0.0	0.0
ESC1	0.0	0.0	0.0	0.0	0.0
ESC2	0.0	0.0	0.0	0.0	0.0
ESC3	0.0	0.0	0.0	0.0	0.0
ESC4	0.0	0.0	0.0	0.0	0.0
ESC5	0.0	0.0	0.0	0.0	0.0
MSC1	0.0	0.0	0.0	0.0	0.0
MSC2	0.0	0.0	0.0	0.0	0.0
MSC3	0.0	0.0	0.0	0.0	0.0
MSC4	0.0	0.0	0.0	0.0	0.0
MSC5	0.0	0.0	0.0	0.0	0.0

	GSC6	ASC1	ASC2	ASC3	ASC4
GSC6	0.482				
ASC1	0.0	0.553			
ASC2	0.0	0.0	0.292		
ASC3	0.0	0.0	0.0	0.219	
ASC4	0.0	0.0	0.0	0.0	0.306
	0.0	0.0	0.0	0.0	0.0
	0.0	0.0	0.0	0.0	0.0
	0.0	0.0	0.0	0.0	0.0
	0.0	0.0	0.0	0.0	0.0
	0.0	0.0	0.0	0.0	0.0
	0.0	0.0	0.0	0.0	0.0
	0.0	0.0	0.0	0.0	0.0
	0.0	0.0	0.0	0.0	0.0
	0.0	0.0	0.0	0.0	0.0
	0.0	0.0	0.0	0.0	0.0
	0.0	0.0	0.0	0.0	0.0

TABLE 4.3. Continued

THETA DELTA

	ASC5	ESC1	ESC2	ESC3	ESC4
ASC5	0.352				
ESC1	0.0	0.650			
ESC2	0.0	0.0	0.560		
ESC3	0.0	0.0	0.0	0.652	
ESC4	0.0	0.0	0.0	0.0	0.552
ESC5	0.0	0.0	0.0	0.0	0.0
MSC1	0.0	0.0	0.0	0.0	0.0
MSC2	0.0	0.0	0.0	0.0	0.0
MSC3	0.0	0.0	0.0	0.0	0.0
MSC4	0.0	0.0	0.0	0.0	0.0
MSC5	0.0	0.0	0.0	0.0	0.0

	ESC5	MSC1	MSC2	MSC3	MSC4
	0.809				
	0.0	0.419			
	0.0	0.0	0.267		
	0.0	0.0	0.0	0.220	
	0.0	0.0	0.0	0.0	0.207
	0.0	0.0	0.0	0.0	0.0

THETA DELTA

	MSC5
MSC5	0.351

STANDARD ERRORS

LAMBDA X

	KSI 1	KSI 2	KSI 3	KSI 4
GSC1	0.0	0.0	0.0	0.0
GSC2	0.049	0.0	0.0	0.0
GSC3	0.046	0.0	0.0	0.0
GSC4	0.049	0.0	0.0	0.0
GSC5	0.045	0.0	0.0	0.0
GSC6	0.049	0.0	0.0	0.0
ASC1	0.0	0.0	0.0	0.0
ASC2	0.0	0.080	0.0	0.0
ASC3	0.0	0.081	0.0	0.0
ASC4	0.0	0.080	0.0	0.0
ASC5	0.0	0.079	0.0	0.0
ESC1	0.0	0.0	0.0	0.0
ESC2	0.0	0.0	0.114	0.0
ESC3	0.0	0.0	0.109	0.0
ESC4	0.0	0.0	0.115	0.0
ESC5	0.0	0.0	0.101	0.0
MSC1	0.0	0.0	0.0	0.0
MSC2	0.0	0.0	0.0	0.058
MSC3	0.0	0.0	0.0	0.057
MSC4	0.0	0.0	0.0	0.057
MSC5	0.0	0.0	0.0	0.058

PHI

	KSI 1	KSI 2	KSI 3	KSI 4
KSI 1	0.065			
KSI 2	0.032	0.057		
KSI 3	0.033	0.029	0.053	
KSI 4	0.034	0.036	0.025	0.062

TABLE 4.3. Continued

THETA DELTA

	GSC1	GSC2	GSC3	GSC4	GSC5
GSC1	0.025				
GSC2	0.0	0.033			
GSC3	0.0	0.0	0.025		
GSC4	0.0	0.0	0.0	0.035	
GSC5	0.0	0.0	0.0	0.0	0.023
GSC6	0.0	0.0	0.0	0.0	0.0
ASC1	0.0	0.0	0.0	0.0	0.0
ASC2	0.0	0.0	0.0	0.0	0.0
ASC3	0.0	0.0	0.0	0.0	0.0
ASC4	0.0	0.0	0.0	0.0	0.0
ASC5	0.0	0.0	0.0	0.0	0.0
ESC1	0.0	0.0	0.0	0.0	0.0
ESC2	0.0	0.0	0.0	0.0	0.0
ESC3	0.0	0.0	0.0	0.0	0.0
ESC4	0.0	0.0	0.0	0.0	0.0
ESC5	0.0	0.0	0.0	0.0	0.0
MSC1	0.0	0.0	0.0	0.0	0.0
MSC2	0.0	0.0	0.0	0.0	0.0
MSC3	0.0	0.0	0.0	0.0	0.0
MSC4	0.0	0.0	0.0	0.0	0.0
MSC5	0.0	0.0	0.0	0.0	0.0

	GSC6	ASC1	ASC2	ASC3	ASC4
	0.035				
	0.0	0.039			
	0.0	0.0	0.025		
	0.0	0.0	0.0	0.021	
	0.0	0.0	0.0	0.0	0.025
	0.0	0.0	0.0	0.0	0.0
	0.0	0.0	0.0	0.0	0.0
	0.0	0.0	0.0	0.0	0.0
	0.0	0.0	0.0	0.0	0.0
	0.0	0.0	0.0	0.0	0.0
	0.0	0.0	0.0	0.0	0.0
	0.0	0.0	0.0	0.0	0.0
	0.0	0.0	0.0	0.0	-0.0
	0.0	0.0	0.0	0.0	0.0

THETA DELTA

	ASC5	ESC1	ESC2	ESC3	ESC4
ASC5	0.028				
ESC1	0.0	0.052			
ESC2	0.0	0.0	0.049		
ESC3	0.0	0.0	0.0	0.052	
ESC4	0.0	0.0	0.0	0.0	0.049
ESC5	0.0	0.0	0.0	0.0	0.0
MSC1	0.0	0.0	0.0	0.0	0.0
MSC2	0.0	0.0	0.0	0.0	0.0
MSC3	0.0	0.0	0.0	0.0	0.0
MSC4	0.0	0.0	0.0	0.0	0.0
MSC5	0.0	0.0	0.0	0.0	0.0

	ESC5	MSC1	MSC2	MSC3	MSC4
	0.058				
	0.0	0.031			
	0.0	0.0	0.022		
	0.0	0.0	0.0	0.020	
	0.0	0.0	0.0	0.0	0.019
	0.0	0.0	0.0	0.0	0.0

THETA DELTA

	MSC5
MSC5	0.027

3.2. Adequacy of the Measurement Model

Let us now examine the squared multiple correlations (R^2) for each of the observed variables (item-pair measurements), and the coefficient of determination for the entire model. These values are presented in Table 4.4.

TABLE 4.4. Selected LISREL Output: R^2 and Coefficient of Determination

```
SQUARED MULTIPLE CORRELATIONS FOR X - VARIABLES

  GSC1          GSC2          GSC3          GSC4          GSC5
      0.708         0.359         0.695         0.525         0.743

  GSC6          ASC1          ASC2          ASC3          ASC4
      0.518         0.447         0.703         0.781         0.694

SQUARED MULTIPLE CORRELATIONS FOR X - VARIABLES

  ASC5          ESC1          ESC2          ESC3          ESC4
      0.648         0.350         0.449         0.345         0.448

  ESC5          MSC1          MSC2          MSC3          MSC4
      0.191         0.581         0.733         0.785         0.793

SQUARED MULTIPLE CORRELATIONS FOR X - VARIABLES

  MSC5
      0.649

TOTAL COEFFICIENT OF DETERMINATION FOR X - VARIABLES IS   0.999
```

Looking first at the R^2 values, we see that, overall, the observed variables do a satisfactory job of measuring their target SC factors. Exceptions to this general finding, however, are the first pair of items measuring academic SC (ASC1) and all item-pairs measuring English SC, which are all less than 5.00. While these values are still reasonable, they are indicative of somewhat less reliability than are the other item-pair measurements.

Considering all item-pairs in combination, we can see by the value of the coefficient of determination that the reliability of the measurement model as a whole was exceptionally high (0.999).

3.3. Goodness-of-Fit of Overall Model

As shown in Table 4.5, the adequacy of the model as a whole in representing the observed data could bear some improvement. This statement is based on the value of the likelihood ratio index (χ^2 (183) = 515.56), the GFI (0.825), and the AGFI (0.780). The RMR (0.046) indicates a marginally good fit between the hypothesized (i.e., restricted) model and the observed data (i.e., unrestricted model).

If, on the other hand, we assess the hypothesized model based on subjective indices of fit, the picture appears somewhat better. For example,

TABLE 4.5. Selected LISREL Output: Goodness-of-Fit of Whole Model

---

MEASURES OF GOODNESS OF FIT FOR THE WHOLE MODEL :

CHI-SQUARE WITH 183 DEGREES OF FREEDOM IS   515.56 (PROB. LEVEL = 0.000)

GOODNESS OF FIT INDEX IS 0.900

ADJUSTED GOODNESS OF FIT INDEX IS 0.874

ROOT MEAN SQUARE RESIDUAL IS   0.045

---

the χ^2/df ratio is 2.82. While this value is not great, it is, nevertheless, within the broad range of acceptable values (see Carmines & McIver, 1981).

Additionally, a null model was estimated in order to calculate the BBI. The likelihood ratio index for this model was χ^2 (210) = 5939.91, yielding a BBI of 0.913; this value represents a marginally good fit to the data.

3.4. Goodness-of-Fit of Individual Model Parameters

Since we now know that the hypothesized model, although not a poor fit to the observed data, is really only marginally good, we now need to locate the area of misfit in the model by examining the fit of individual parameters; due to space limitations, only the *T*-values and MIs are included here. This portion of the output is presented in Table 4.6.

A review of the *T*-values reveals all parameter estimates to be substantial; the magnitude of all estimates is >2.00, indicating that they are statistically significant and thus essential to the model.

Looking at the MIs, however, we see that there are six estimates >5.00 in the factor loading matrix, and 25 >5.00 in the error variance/covariance matrix. From a practical perspective, given that we are investigating a single measuring instrument, this finding is not totally unexpected. With respect to the LX matrix, the MIs reflect, for example, that some items are tapping nontarget, as well as target SC factors; considering the known moderate correlation among the four SC factors under study here, the finding should not be surprising. Likewise, with respect to the TD matrix, the MIs indicate correlated measurement errors—again, not an uncommon finding among subscales of the same measuring instrument. Such covariation frequently results from random error introduced by a particular measurement method; one example is that of method effects derived from the item format associated with subscales of the same measuring instrument.

Despite the likelihood that these explanations do account for the numerous MIs associated with the present model, it seems appropriate to investigate the extent to which the MI parameters, if relaxed, lead to an improvement in model fit. One way of determining this information is to investigate, under alternate model specifications, substantial changes to

TABLE 4.6. Selected LISREL Output: *T*-Values and Modification Indices

MODIFICATION INDICES

LAMBDA X

	KSI 1	KSI 2	KSI 3	KSI 4
GSC1	0.0	1.442	0.892	0.789
GSC2	0.0	0.055	1.347	0.218
GSC3	0.0	1.291	0.042	0.002
GSC4	0.0	8.317	0.168	0.431
GSC5	0.0	6.751	0.000	1.661
GSC6	0.0	1.090	0.043	0.201
ASC1	0.692	0.0	2.874	0.346
ASC2	3.209	0.0	0.579	3.192
ASC3	1.592	0.0	5.407	1.949
ASC4	4.856	0.0	17.275	0.718
ASC5	0.354	0.0	0.483	10.206
ESC1	5.684	0.395	0.0	8.522
ESC2	1.703	1.020	0.0	0.006
ESC3	1.921	0.062	0.0	0.078
ESC4	4.836	1.081	0.0	4.368
ESC5	0.003	0.223	0.0	1.651
MSC1	0.000	15.891	3.177	0.0
MSC2	0.347	36.944	12.125	0.0
MSC3	0.115	0.332	0.083	0.0
MSC4	0.347	0.330	0.437	0.0
MSC5	1.065	4.321	1.493	0.0

PHI

	KSI 1	KSI 2	KSI 3	KSI 4
KSI 1	0.0			
KSI 2	0.0	0.0		
KSI 3	0.0	0.0	0.0	
KSI 4	0.0	0.0	0.0	0.0

THETA DELTA

	GSC1	GSC2	GSC3	GSC4	GSC5
GSC1	0.0				
GSC2	0.004	0.0			
GSC3	8.741	0.240	0.0		
GSC4	0.472	2.860	19.802	0.0	
GSC5	9.282	1.764	1.843	8.166	0.0
GSC6	0.096	0.189	1.914	15.671	2.560
ASC1	0.698	0.654	3.793	1.380	0.845
ASC2	0.018	0.000	2.177	2.582	0.125
ASC3	0.269	0.032	1.010	0.555	0.781
ASC4	0.123	0.011	0.457	0.283	0.700
ASC5	0.829	0.019	0.424	0.328	2.229
ESC1	4.356	5.876	0.004	4.932	0.255
ESC2	0.355	2.315	2.321	2.274	1.488
ESC3	2.192	0.048	4.791	1.665	0.006
ESC4	0.133	0.104	4.052	4.423	0.526
ESC5	0.376	0.823	7.989	12.305	0.105
MSC1	0.004	0.343	3.275	9.706	0.014
MSC2	0.003	1.305	2.193	0.207	1.679
MSC3	0.021	0.190	0.194	5.141	7.194
MSC4	0.181	0.509	1.498	0.900	2.179
MSC5	0.004	0.004	0.080	0.295	5.700

GSC6	ASC1	ASC2	ASC3	ASC4
0.0				
5.140	0.0			
0.634	31.380	0.0		
0.181	2.851	13.611	0.0	
1.397	0.042	4.346	4.670	0.0
6.048	10.572	17.952	4.611	2.837
0.588	0.009	0.312	0.704	0.137
10.310	0.298	1.490	0.581	1.023
4.646	0.828	0.002	1.823	0.631
0.008	1.328	2.426	3.779	11.603
0.087	0.361	0.006	0.007	0.019
0.647	23.291	15.599	5.228	6.843
4.437	0.398	16.581	1.662	0.139
0.972	7.598	3.959	4.109	2.157
4.127	2.492	2.758	0.058	0.722
2.058	1.436	0.499	1.040	2.703

TABLE 4.6. Continued

THETA DELTA

	ASC5	ESC1	ESC2	ESC3	ESC4
ASC5	0.0				
ESC1	3.202	0.0			
ESC2	4.733	2.551	0.0		
ESC3	3.428	1.703	0.181	0.0	
ESC4	0.441	0.529	0.144	1.595	0.0
ESC5	0.005	0.003	9.992	3.931	1.679
MSC1	14.370	0.532	0.094	0.753	0.467
MSC2	0.502	3.223	0.827	0.025	0.940
MSC3	0.529	2.000	4.633	0.154	1.655
MSC4	14.567	0.183	0.063	0.527	0.022
MSC5	2.143	0.040	1.178	0.015	3.526

	ESC5	MSC1	MSC2	MSC3	MSC4
	0.0				
	4.327	0.0			
	0.002	1.008	0.0		
	2.370	2.427	3.404	0.0	
	1.214	14.351	1.058	0.322	0.0
	1.202	2.870	0.838	19.314	1.953

THETA DELTA

	MSC5
MSC5	0.0

MAXIMUM MODIFICATION INDEX IS   35.94 FOR ELEMENT (18, 2) OF LAMBDA X

T-VALUES

LAMBDA X

	KSI 1	KSI 2	KSI 3	KSI 4
GSC1	0.0	0.0	0.0	0.0
GSC2	18.299	0.0	0.0	0.0
GSC3	21.504	0.0	0.0	0.0
GSC4	17.500	0.0	0.0	0.0
GSC5	22.622	0.0	0.0	0.0
GSC6	17.346	0.0	0.0	0.0
ASC1	0.0	0.0	0.0	0.0
ASC2	0.0	15.748	0.0	0.0
ASC3	0.0	16.370	0.0	0.0
ASC4	0.0	15.618	0.0	0.0
ASC5	0.0	15.135	0.0	0.0
ESC1	0.0	0.0	0.0	0.0
ESC2	0.0	0.0	9.813	0.0
ESC3	0.0	0.0	9.155	0.0
ESC4	0.0	0.0	9.864	0.0
ESC5	0.0	0.0	7.334	0.0
MSC1	0.0	0.0	0.0	0.0
MSC2	0.0	0.0	0.0	19.513
MSC3	0.0	0.0	0.0	20.241
MSC4	0.0	0.0	0.0	20.444
MSC5	0.0	0.0	0.0	18.137

PHI

	KSI 1	KSI 2	KSI 3	KSI 4
KSI 1	10.829			
KSI 2	5.879	7.320		
KSI 3	5.820	6.346	5.027	
KSI 4	4.305	8.677	2.665	9.404

TABLE 4.6. Continued

THETA DELTA

	GSC1	GSC2	GSC3	GSC4	GSC5
GSC1	11.724				
GSC2	0.0	13.339			
GSC3	0.0	0.0	11.923		
GSC4	0.0	0.0	0.0	13.565	
GSC5	0.0	0.0	0.0	0.0	11.088
GSC6	0.0	0.0	0.0	0.0	0.0
ASC1	0.0	0.0	0.0	0.0	0.0
ASC2	0.0	0.0	0.0	0.0	0.0
ASC3	0.0	0.0	0.0	0.0	0.0
ASC4	0.0	0.0	0.0	0.0	0.0
ASC5	0.0	0.0	0.0	0.0	0.0
ESC1	0.0	0.0	0.0	0.0	0.0
ESC2	0.0	0.0	0.0	0.0	0.0
ESC3	0.0	0.0	0.0	0.0	0.0
ESC4	0.0	0.0	0.0	0.0	0.0
ESC5	0.0	0.0	0.0	0.0	0.0
MSC1	0.0	0.0	0.0	0.0	0.0
MSC2	0.0	0.0	0.0	0.0	0.0
MSC3	0.0	0.0	0.0	0.0	0.0
MSC4	0.0	0.0	0.0	0.0	0.0
MSC5	0.0	0.0	0.0	0.0	0.0

	GSC6	ASC1	ASC2	ASC3	ASC4
GSC6	13.605				
ASC1	0.0	14.025			
ASC2	0.0	0.0	11.876		
ASC3	0.0	0.0	0.0	10.356	
ASC4	0.0	0.0	0.0	0.0	12.092
ASC5	0.0	0.0	0.0	0.0	0.0
ESC1	0.0	0.0	0.0	0.0	0.0
ESC2	0.0	0.0	0.0	0.0	0.0
ESC3	0.0	0.0	0.0	0.0	0.0
ESC4	0.0	0.0	0.0	0.0	0.0
ESC5	0.0	0.0	0.0	0.0	0.0
MSC1	0.0	0.0	0.0	0.0	0.0
MSC2	0.0	0.0	0.0	0.0	0.0
MSC3	0.0	0.0	0.0	0.0	0.0
MSC4	0.0	0.0	0.0	0.0	0.0
MSC5	0.0	0.0	0.0	0.0	0.0

THETA DELTA

	ASC5	ESC1	ESC2	ESC3	ESC4
ASC5	12.659				
ESC1	0.0	12.549			
ESC2	0.0	0.0	11.365		
ESC3	0.0	0.0	0.0	12.563	
ESC4	0.0	0.0	0.0	0.0	11.243
ESC5	0.0	0.0	0.0	0.0	0.0
MSC1	0.0	0.0	0.0	0.0	0.0
MSC2	0.0	0.0	0.0	0.0	0.0
MSC3	0.0	0.0	0.0	0.0	0.0
MSC4	0.0	0.0	0.0	0.0	0.0
MSC5	0.0	0.0	0.0	0.0	0.0

	ESC5	MSC1	MSC2	MSC3	MSC4
ESC5	13.975				
MSC1	0.0	13.547			
MSC2	0.0	0.0	12.003		
MSC3	0.0	0.0	0.0	11.095	
MSC4	0.0	0.0	0.0	0.0	10.766
MSC5	0.0	0.0	0.0	0.0	0.0

THETA DELTA

	MSC5
MSC5	13.029

major parameters (λs and ϕs) in the originally hypothesized model. Post hoc model fitting of this nature has been termed "sensitivity analysis" (Byrne et al., 1989).

4. Post Hoc Analyses

The reader is reminded of the caveats presented in Chapter 3 concerning the conduct of post hoc analyses. As noted earlier, such analyses are of an exploratory nature and should be subjected to cross-validation with independent samples before drawing firm conclusions from their findings. As long as the researcher is cognizant of this fact, the conduct of post hoc analyses can provide a valuable insight into the model under study. One way to think of the post hoc process is as a "sensitivity analysis" whereby practical, as well as statistical significance, are taken into account. Let's now turn to an application of these procedures to the present data.

4.1. Model-Fitting Procedures

Using the MIs as the primary guide, a series of nested alternative models were respecified and reestimated beyond the initially fitted model. In total, 27 post hoc models were specified; these included 22 error covariances and 5 secondary loadings (item-pair loadings on nontarget factors). The model-fitting process was continued until a statistically nonsignificant model was reached. This final model yielded a χ^2 likelihood ratio index of 183.75 ($p = 0.06$); the BBI was 0.969).

4.2. Sensitivity Analyses

Although we have determined the model of best fit statistically, the question now focuses on the practical significance of these additional parameters (i.e., their importance to the overall meaningfulness of the model). Given the known sensitivity of the χ^2 statistic to sample size, there is always the concern of overfitting the model; that is, fitting the model to trivial sample-specific artifacts in the data.

One way of determining this information is to test the sensitivity of major parameters in the model to the addition of the post hoc parameters. For example, if the estimates of major parameters undergo no appreciable change when minor parameters are added to the model, this is an indication that the initially hypothesized model is empirically robust; the more fitted model therefore represents a minor improvement to an already adequate model and the additional parameters should be deleted from the model. If, on the other hand, the major parameters undergo substantial alteration, the exclusion of the post hoc parameters may lead to biased estimates (Alwin & Jackson, 1980; Jöreskog, 1983); the minor parameters should therefore be retained in the model.

One method of estimating this information is to correlate major parameters (the λs and φs) in the baseline model with those in the best-fitting post hoc model; this is easily performed by means of SPSSX or any other similar computer package. Coefficients close to 1.00 argue for the stability of the initial model, and thus the triviality of the minor parameters in the post hoc model. In contrast, coefficients that are not close to 1.00 (say, <0.90) are an indication that the major parameters are adversely affected, and thus argues for inclusion of the post hoc parameters in the final model.

This suggestion, however, is intended to serve only as a general guide to post hoc model fitting. Clearly, decisions regarding the inclusion or exclusion of post hoc parameters must involve the weighing of many additional factors. Other considerations, for example, might include: the magnitude (mean or median) of secondary loading and/or error covariance estimates (values less than 1.5 are considered trivial); differences in incremental fit based on subjective indices, rather than on the χ^2 statistic (see e.g., Marsh & Hocevar, 1985); and the substantive meaningfulness of the model relative to the theory and other empirical research in the area.

Applied to the present data, the preceding analyses yielded the following information: the correlation between the estimated factor loadings of the initial and final models was 0.997; the correlation between the estimated variance/covariances of the initial and final models was 0.988; the error covariance estimates of the final model, while statistically significant, ranged from 0.02 to 0.14 (Md = 0.06, which was considered to be relatively minor); the estimated secondary factor loadings, while statistically significant were also relatively minor, ranging from 0.12 to 0.24 (Md = 0.07); and there were no incremental differences between the BBI values (Δ BBI = 0.000). This information, together with the fact that, substantively, these artifacts in the data are not unreasonable when analyses are focused on a single measuring instrument, led to the rejection of the final post hoc model in favor of the more parsimonious initially hypothesized model as presented in Figure 4.1 and Table 4.1.

In light of these findings, it was concluded that the SDQIII is a psychometrically sound instrument for measuring multidimensional facets of adolescent SC.

5. Summary

This chapter focused on the application of LISREL CFA modeling in the validation of a measuring instrument. We examined, in detail, the LISREL input for this model and selected portions of the output. Of particular importance in this chapter was the conduct of a sensitivity analysis, the crucial component of post hoc model-fitting procedures, to determine the practical significance of additional parameters to the model.

5
Validating Multiple Traits Assessed by Multiple Methods: The Multitrait-Multimethod Framework

In this application, we use LISREL CFA procedures to model hypotheses related to convergent and discriminant validity, two indicators of construct validity (see Campbell & Fiske, 1959). Specifically, we use a LISREL modeling approach to examine data within a multitrait-multimethod (MTMM) framework. As such, our analyses now focus on the validity of multiple traits assessed by multiple methods. (For details of the study related to this application, see Byrne, in press.)

The multiple traits in the present application, once again, are the four facets of SC: general SC, academic SC, English SC, and mathematics SC. The multiple methods are three different scaling techniques: Likert, semantic differential, and Guttman scales, as represented by the SDQIII, API, and the SES and SCA, respectively. Although the reference cited for this application examines construct validity for both low and high academically tracked high school students, we will focus on analyses related to the low track only; tests for invariance will be addressed in Chapter 6.

1. Assessment of Construct Validity: The MTMM Matrix

Campbell and Fiske (1959) posited that claims of construct validity must be accompanied by evidence of both convergent and discriminant validity. As such, a measure should correlate highly with other measures to which it is theoretically linked (convergent validity) and negligibly with those that are theoretically unrelated (discriminant validity). To determine evidence of construct validity, Campbell and Fiske proposed that measures of multiple traits be assessed by multiple methods and that all trait-method correlations be arranged in an MTMM matrix.

1.1. The Campbell-Fiske Approach to MTMM Analyses

The assessment of construct validity then focuses on comparisons among three blocks of correlations: scores on the same traits measured by different methods (monotrait-heteromethod values, i.e., convergent validity), scores on different traits measured by the same method (heterotrait-monomethod values, i.e., discriminant validity), and scores on different traits measured by different methods (heterotrait-heteromethod values, i.e., discriminant validity). Specific criteria guide the inspection of these values, but will not be discussed here since they are not relevant to the present analyses. However, interested readers are referred to Byrne (in press) for an extensive discussion of this technique.

Recently, methodologists have argued that a more sophisticated approach to assessing construct validity within the MTMM framework is the analysis of covariance structures using LISREL modeling procedures. Indeed, the CFA approach has been shown to have several advantages over the Campbell-Fiske approach (see Marsh & Hocevar, 1983; Schmitt & Stults, 1986; Widaman, 1985). First the MTMM matrix is explained in terms of the underlying latent constructs, rather than the observed variables. Second the evaluation of convergent and discriminant validities can be made at both the matrix and individual parameter levels. Third, hypotheses related to convergent and discriminant validity can be tested statistically, based on a series of hierarchically nested models. Finally, separate estimates of variance due to traits, methods, and uniqueness are provided.

1.2. The LISREL Approach to MTMM Analyses

The first step in analyzing data within an MTMM framework is to formulate a LISREL model that comprises both the trait and method factors. In the present application, we have four traits and three methods, yielding a seven-factor model. A schematic presentation of this model is illustrated in Figure 5.1 and the related pattern of parameters is shown in Table 5.1.

Compared with our two previous applications, we can see that this model represents a more complex structure; three aspects of the parameter specifications are important to note. First, each observed variable (the Xs) loads on two factors—a trait as well as a method factor. Second, in contrast with our two previous applications, the first λ of each set of congeneric measures in the factor loading matrix (Λ) is not fixed to 1.0 for purposes of identification. Alternatively, each trait (ϕ_{11} to ϕ_{44}) and method (ϕ_{55} to ϕ_{77}) factor variance has been fixed to 1.0 for the same purpose. Third, as recommended (Schmitt & Stults, 1986; Widaman, 1985), the trait-method factors[1] have been fixed to 0.0 to alleviate prob-

---

[1] The trait-method correlations are represented by: PH(5,1), PH(6,1), PH(7,1), PH(5,2), PH(6,2), PH(7,2), PH(5,3), PH(6,3), PH(7,3), PH(5,4), PH(6,4), and PH(7,4).

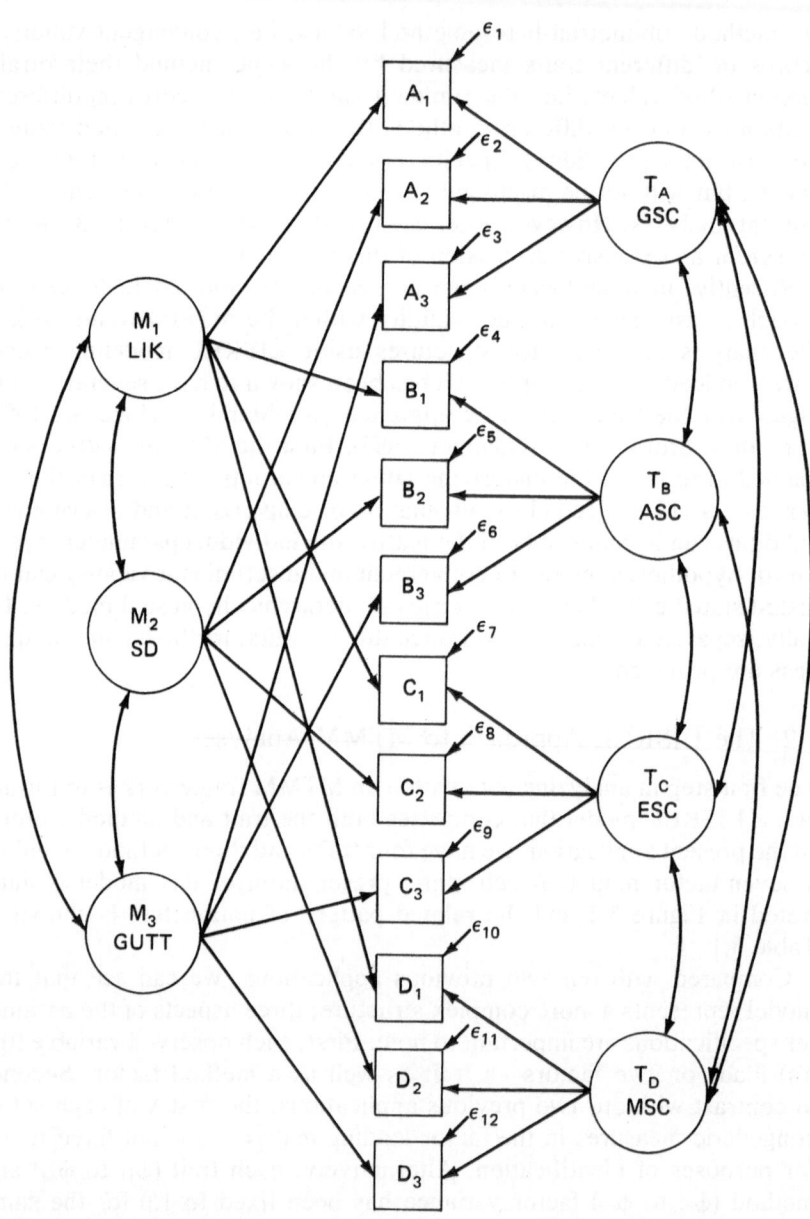

FIGURE 5.1. Multitrait-Multimethod Model of Data: M = method; T = trait; LIK = Likert scale; SD = semantic differential scale; GUTT = Guttman scale, GSC = general self-concept; ASC = academic self-concept; ESC = English self-concept; and MSC = mathematics self-concept. From Byrne (in press), "Multigroup Comparisons and the Assumption of Equivalent Construct Validity Across Groups: Methodological and Substantive Issues" in *Multivariate Behavioral Research.* Copyright 1989 by Lawrence Erlbaum Associates, Inc. Reprinted with permission.

TABLE 5.1. Pattern of Estimated Parameters for Hypothesized Seven-Factor CFA MTMM Model

Factor Loading Matrix (Λ_x)

Measure	X	Traits				Methods		
		GSC (ξ_1)	ASC (ξ_2)	ESC (ξ_3)	MSC (ξ_4)	LIK (ξ_5)	SD (ξ_6)	GUTT (ξ_7)
SDQGSC	1	λ_{11}	.0	.0	.0	λ_{15}	.0	.0
SDQASC	2	.0	λ_{22}	.0	.0	λ_{25}	.0	.0
SDQESC	3	.0	.0	λ_{33}	.0	λ_{35}	.0	.0
SDQMSC	4	.0	.0	.0	λ_{44}	λ_{45}	.0	.0
APIGSC	5	λ_{51}	.0	.0	.0	.0	λ_{56}	.0
APIASC	6	.0	λ_{62}	.0	.0	.0	λ_{66}	.0
APIESC	7	.0	.0	λ_{73}	.0	.0	λ_{76}	.0
APIMSC	8	.0	.0	.0	λ_{84}	.0	λ_{86}	.0
SESGSC	9	λ_{91}	.0	.0	.0	.0	.0	λ_{97}
SCAASC	10	.0	$\lambda_{10,2}$	.0	.0	.0	.0	$\lambda_{10,7}$
SCAESC	11	.0	.0	$\lambda_{11,3}$	.0	.0	.0	$\lambda_{11,7}$
SCAMSC	12	.0	.0	.0	$\lambda_{12,4}$	.0	.0	$\lambda_{12,7}$

Factor Variance-Covariance Matrix (Φ)

	Traits				Methods		
	GSC (ϕ_{11})	ASC (ϕ_{22})	ESC (ϕ_{33})	MSC (ϕ_{44})	LIK (ϕ_{55})	SD (ϕ_{66})	GUTT (ϕ_{77})
GSC	ϕ_{11}						
ASC	ϕ_{21}	ϕ_{22}					
ESC	ϕ_{31}	ϕ_{32}	ϕ_{33}				
MSC	ϕ_{41}	ϕ_{42}	ϕ_{43}	ϕ_{44}			
LIK	.0	.0	.0	.0	ϕ_{55}		
SD	.0	.0	.0	.0	ϕ_{65}	ϕ_{66}	
GUTT	.0	.0	.0	.0	ϕ_{75}	ϕ_{76}	ϕ_{77}

TABLE 5.1. Continued

Error Variance-Covariance Matrix (Θ_δ)

	X_1	X_2	X_3	X_4	X_5	X_6	X_7	X_8	X_9	X_{10}	X_{11}	X_{12}
X_1	$\Theta\delta_1$	.0	.0	.0	.0	.0	.0	.0	.0	.0	.0	.0
X_2	.0	$\Theta\delta_2$	.0	.0	.0	.0	.0	.0	.0	.0	.0	.0
X_3	.0	.0	$\Theta\delta_3$	.0	.0	.0	.0	.0	.0	.0	.0	.0
X4	.0	.0	.0	$\Theta\delta_4$	.0	.0	.0	.0	.0	.0	.0	.0
X_5	.0	.0	.0	.0	$\Theta\delta_5$	.0	.0	.0	.0	.0	.0	.0
X_6	.0	.0	.0	.0	.0	$\Theta\delta_6$	.0	.0	.0	.0	.0	.0
X_7	.0	.0	.0	.0	.0	.0	$\Theta\delta_7$	.0	.0	.0	.0	.0
X_8	.0	.0	.0	.0	.0	.0	.0	$\Theta\delta_8$	.0	.0	.0	.0
X_9	.0	.0	.0	.0	.0	.0	.0	.0	$\Theta\delta_9$	.0	.0	.0
X_{10}	.0	.0	.0	.0	.0	.0	.0	.0	.0	$\Theta\delta_{10}$	.0	.0
X_{11}	.0	.0	.0	.0	.0	.0	.0	.0	.0	.0	$\Theta\delta_{11}$	.0
X_{12}	.0	.0	.0	.0	.0	.0	.0	.0	.0	.0	.0	$\Theta\delta_{12}$

lems related to identification and estimation. Finally, both the traits and the methods are specified as being correlated among themselves.[2]

To test this seven-factor model for evidence of convergent and discriminant validity, we compare it against a series of more restrictive models in which specific parameters have either been eliminated or constrained to equal zero. Since the difference in $\chi^2(\Delta\chi)^2$ is itself χ^2-distributed with degrees of freedom equal to the difference in degrees of freedom for the two models, the fit differential between comparison models can be tested statistically. A significant $\Delta\chi^2$ argues for the superiority of the less restrictive model. (For a more extensive discussion of these alternative models and their comparisons, see Widaman, 1985.)

A total of eight models, in addition to the hypothesized seven-factor model, are specified in the present application; five are used to make comparisons related to convergent and discriminant validity and three to determine the degree of method bias associated with each scaling method. Before examining these model comparisons, however, let us first study the LISREL input and output for the proposed seven-factor model.

---

[2]The trait correlations are represented by PH(2,1), PH(3,1), PH(4,1), PH(3,2), PH(4,2), and PH(4,3); the method correlations are represented by PH(6,5), PH(7,5), and PH(7,6).

2. LISREL Input

Translation of the pattern of parameter estimates into LISREL input statements for this model (see Table 5.1) is presented in Table 5.2; the data correlation matrix of means and standard deviations is also included.

We begin by reviewing this MTMM model specification in light of our LISREL input statements.

2.1. The DA Card

Although the DA card tells us that there are 15 variables (NI = 15), only 12 are used in the present application; thus, the SE card is used. (Recall from Chapter 3 that the other three variables, academic scores, are not included in these analyses). We see also that the sample size is 252 (NO

TABLE 5.2. LISREL Input and Parameter Specifications of Hypothesized Seven-Factor MTMM Model

```
MTMM-G-LEVEL(MODEL1D)-TRAITS AND METHODS CORRELATED (WIDMN#3C)-"MTMMG1D"
DA NI=15 NO=252 MA=KM
LA
*
'SDQGSC' 'SDQASC' 'SDQESC' 'SDQMSC' 'APIGSC' 'SESGSC' 'APIASC' 'SCAASC'
'APIESC' 'SCAESC' 'APIMSC' 'SCAMSC' 'GPA' 'ENG' 'MATH'
KM SY
(15F4.3)
1000
 3211000
 301 2801000
 243 354-0601000
 610 236 205 2771000
 752 263 270 258 5861000
 452 567 379 351 554 4571000
 275 582 252 232 229 273 5241000
 146 427 518 032 184 110 474 3741000
 236 372 430 004 262 246 413 507 4981000
 263 389 054 778 260 236 421 350 233 0241000
 237 350-019 720 211 216 372 446 078 075 7461000
 038 361 027 125 049 009 342 452 063 256 035 1541000
 059 320 078 103 135 038 318 252 144 435-020 010 6971000
 009 232 014 344 036-021 199 355 065 086 280 490 647 3421000
ME
*
76.004 49.533 54.921 41.591 76.881 31.193 70.290 24.802 57.821 25.333
41.831 23.020 70.440 68.737 62.637
SD
*
13.400 12.396 9.448 13.356 9.078 4.343 8.345 4.474 10.627 4.343 10.609
6.319 10.172 11.738 16.203
SELECTION
1 2 3 4 5 7 9 11 6 8 10 12/
MO NX=12 NK=7 LX=FU PH=SY,FI TD=DI
FR LX(1,1) LX(5,1) LX(9,1) LX(2,2) LX(6,2) LX(10,2) LX(3,3) LX(7,3)
FR LX(11,3) LX(4,4) LX(8,4) LX(12,4)
FR LX(1,5) LX(2,5) LX(3,5) LX(4,5) LX(5,5) LX(6,6) LX(7,6) LX(8,6)
FR LX(9,7) LX(10,7) LX(11,7) LX(12,7)
FR PH(2,1) PH(3,1) PH(4,1) PH(3,2) PH(4,2) PH(4,3)
FR PH(6,5) PH(7,5) PH(7,6)
ST 1.00 PH(1,1) PH(2,2) PH(3,3) PH(4,4) PH(5,5) PH(6,6) PH(7,7)
ST .9 LX(1,1) LX(5,1) LX(9,1) LX(3,3) LX(7,3)
ST -.4 LX(2,2) LX(6,2) LX(10,2)
ST .9 LX(11,3) LX(4,4) LX(8,4) LX(12,4)
ST .9 LX(1,5) LX(2,5) LX(3,5) LX(4,5) LX(5,6) LX(6,6) LX(7,6) LX(8,6)
ST .9 LX(9,7) LX(10,7) LX(11,7) LX(12,7)
ST .02 PH(2,1) PH(3,1) PH(4,1) PH(3,2) PH(4,2) PH(4,3)
ST .02 PH(6,5) PH(7,5) PH(7,6)
ST .2 TD(1,1) TD(2,2) TD(3,3) TD(4,4) TD(5,5) TD(6,6) TD(7,7) TD(8,8)
ST .2 TD(9,9) TD(10,10) TD(11,11) TD(12,12)
OU NS SE TV MI
```

TABLE 5.2. Continued

PARAMETER SPECIFICATIONS

LAMBDA X

	KSI 1	KSI 2	KSI 3	KSI 4	KSI 5
SDJGSC	1	0	0	0	2
SDJASC	0	3	0	0	4
SDQESC	0	0	5	0	6
SDJMSC	0	0	0	7	8
APIGSC	9	0	0	0	0
APIASC	0	11	0	0	0
APIESC	0	0	13	0	0
APIMSC	0	0	0	15	0
SESJSC	17	0	0	0	0
SCAASC	0	19	0	0	0
SCAESC	0	0	21	0	0
SCAMSC	0	0	0	23	0

KSI 6	KSI 7
0	0
0	0
0	0
0	0
10	0
12	0
14	0
16	0
0	18
0	20
0	22
0	24

PHI

	KSI 1	KSI 2	KSI 3	KSI 4	KSI 5
KSI 1	0				
KSI 2	25	0			
KSI 3	26	27	0		
KSI 4	28	29	30	0	
KSI 5	0	0	0	0	0
KSI 6	0	0	0	0	31
KSI 7	0	0	0	0	32

KSI 6	KSI 7
0	
33	0

THETA DELTA

SDJGSC 34	SDJASC 35	SDQESC 36	SDQMSC 37	APIGSC 38
APIASC 39	APIESC 40	APIMSC 41	SESGSC 42	SCAASC 43

THETA DELTA

SCAESC 44	SCAMSC 45

= 252) and that the data are to be analyzed as a correlation matrix (MA = KM). Finally, the input data are in the form of a symmetric correlation matrix (KM SY).

2.2. The MO Card

The MO card tells us that we are working with an all-X model that consists of 12 X variables (NX = 12) and 7 latent (ξ) factors (NK = 7); the factor loading matrix is full and fixed (LX = FU); the factor variance/covariance matrix is symmetric and free (PH = SY); and the error variance/covariance matrix is diagonal and free (TD = DI).[3] In Chapter 4, it was noted that specification of the error matrix as TD = SY, FI, can act as a time-saver in the event that the researcher wishes to conduct post hoc analyses that include correlated errors. The error matrix has not been specified in this way here because in MTMM applications, model fitting beyond the initially hypothesized model leads to innumerable problems related to estimation and other considerations; thus, post hoc model fitting is not recommended (Widaman, 1985).

2.3. The OU Card

The OU Card has specified that no start values are to be provided by the program (NS), and that standard errors (SE), T-values (TV), and modification indices (MI) are to be printed in the output.

3. LISREL Output

The assessment of convergent and discriminant validity can be determined in two major ways: examination of individual parameters representing trait and method factors, and the comparison of MTMM models. Since the model comparison approach requires the input of a series of alternatively specified models, we shall leave this procedure until later, and we turn now to the the evaluation of individual parameters. We first examine, however, the goodness of fit for the overall hypothesized seven-factor model.

3.1. Goodness-of-Fit of the Overall Model

The overall fit of this model, based on the χ^2 likelihood ratio index, was slightly less than optimal ($\chi^2_{(33)} = 105.21$), indicating some degree of misfit in the model. However, as noted earlier, post hoc model fitting of MTMM

---

[3]Recall that PH = SY and TD = DI represent free matrices by default.

TABLE 5.3. Selected LISREL Output: Goodness of Fit of Whole Model

```
    MEASURES OF GOODNESS OF FIT FOR THE WHOLE MODEL :

CHI-SQUARE WITH  33 DEGREES OF FREEDOM IS      105.21
(PROB. LEVEL = 0.000)

        GOODNESS OF FIT INDEX IS 0.879

    ADJUSTED GOODNESS OF FIT INDEX IS 0.715

      ROOT MEAN SQUARE RESIDUAL IS       0.045
```

models is problematic and not recommended.[4] Goodness-of-fit values for the whole model are presented in Table 5.3.

3.2. Evidence of Convergent and Discriminant Validities

Assessments of trait- and method-related variance can be ascertained by examining individual parameter estimates; we now focus on this mode of determining evidence of convergent and discriminant validity. These values, together with the standard errors, are summarized in Table 5.4; statistical significance, as determined by the T-values, is indicated by means of asterisks.

The magnitude of the trait loadings provides an indication of convergent validity. As indicated in Table 5.4, all loadings were substantial and statistically significant. We can conclude from these results that each trait factor was well defined by the hypothesized model.

Except for the measurement of academic SC by the Guttman scale ($\lambda_{10,7}$ = 0.73), method factors tended to have weak to moderate loadings; only 7 of the 12 parameters were significant. These results indicate that method bias effects, while present to a moderate degree for each of the scales, was really only problematic in the measurement of academic SC by the SCA.

Discriminant validity of traits and methods are determined by examining their respective factor correlation matrices. Looking first at trait discrimination, we see that, except for correlations between English and academic SC (0.72) and between mathematics and English SC (0.08), only a modest claim of discriminant validity can be made. This finding, how-

---

[4]Since the largest MI pointed to an error covariance between the APIESC and APIMSC, one further model was estimated in which TD(87) was specified as free. However, as predicted by Widaman, problems of convergence curtailed the estimation of parameters.

TABLE 5.4. Factor and Error/Uniqueness Loadings and Factor Correlations for Baseline Model-Low Track[a]

Measure	Trait				Method			Error/uniqueness
	1	2	3	4	I	II	III	
Likert								
General SC	.89* (.05)	.0	.0	.0	.07 (.07)	.0	.0	.20* (.05)
Academic SC	.0	.73* (.06)	.0	.0	.31* (.11)	.0	.0	.37* (.07)
English SC	.0	.0	.78* (.07)	.0	.41* (.16)	.0	.0	.22 (.16)
Mathematics SC	.0	.0	.0	.87* (.05)	.08 (.06)	.0	.0	.24* (.03)
Semantic Differential								
General SC	.67* (.06)	.0	.0	.0	.0	.46* (.16)	.0	.32* (.15)
Academic SC	.0	.77* (.06)	.0	.0	.0	.43* (.15)	.0	.21 (.11)
English SC	.0	.0	.78* (.06)	.0	.0	.02 (.07)	.0	.37* (.06)
Mathematics SC	.0	.0	.0	.88* (.05)	.0	.05 (.05)	.0	.21* (.03)
Guttman								
General SC	.84* (.06)	.0	.0	.0	.0	.0	.01 (.05)	.30* (.05)
Academic SC	.0	.65* (.06)	.0	.0	.0	.0	.73* (.13)	.04 (.17)
English SC	.0	.0	.63* (.06)	.0	.0	.0	.27* (.07)	.53* (.06)
Mathematics SC	.0	.0	.0	.84* (.05)	.0	.0	.24* (.06)	.25* (.04)

Factor correlations							
Trait 1	1.0						
Trait 2	.59 (.06)	1.0					
Trait 3	.33* (.07)	.72* (.05)	1.0				
Trait 4	.34* (.06)	.52* (.06)	.08 (.07)	1.0			
Method I	.0	.0	.0	.0	1.0		
Method II	.0	.0	.0	.0	.11 (.17)	1.0	
Method III	.0	.0	.0	.0	.39* (.13)	.03 (.12)	1.0

[a] All values of 1.0 and .0 are fixed values. All parameter estimates differing significantly from zero have asterisks. Parenthesized values are standard errors of associated parameters; SC = self-concept.
From Byrne (in press), "Multigroup Comparisons and the Assumption of Equivalent Construct Validity Across Groups: Methodological and Substantive Issues" in Multivariate Behavioral Research. Copyright 1989 by Lawrence Erlbaum Associates, Inc. Reprinted with permission.

ever, is consistent with current SC literature. In this regard, Marsh and Hocevar (1983) have noted that only when correlations are extreme (i.e., approach unity) should researchers be concerned about a lack of discriminant validity. As such, claims of discriminant validity of the traits in the present data appear justified. Additionally, Marsh and Hocevar have argued that the trait correlations should be consistent with the underlying theory. In this regard, the trait correlations are consistent with the Shavelson et al. (1976) hierarchical model of SC and with other empirical findings in the SC research (see e.g., Marsh, Byrne, & Shavelson, 1988).

From the results presented in Table 5.4 we can conclude that discriminant validity among the method factors was reasonably good. These findings suggest that, for the most part, method effects associated with each measurement scale were fairly independent of the other measurement methods incorporated in the model.

4. Comparison of MTMM Models

We turn our attention now to the assessment of convergent and discriminant validity by means of comparisons between pairs of alternatively specified models. Of particular interest are comparisons between the hypothesized seven-factor model and other more restrictive models. The seven-factor model serves as the baseline model since it represents hypothesized relations among the traits and methods, and typically demonstrates the best fit to the data; it is the least restrictive model, having both correlated traits and correlated methods. The models presented here follow from the work of Widaman (1985).

We will first examine the LISREL input for each comparison model; only the statements beginning with the MO card will be included here since the first six cards remain the same for all models. To assist the reader in conceptualizing the pattern of model parameters, the LISREL numbered summary of parameter specifications will accompany each model input. This information will be followed by a summary of goodness-of-fit statistics for each model and the results of all model comparisons.

4.1. LISREL Input for Comparison Models

Model 1 (see Table 5.5) represents the null model and, as such, is the most restrictive of all the models. It hypothesizes that each observed variable is an independent factor (NX = 12; NK = 12); such independence thus precludes correlations among the factors (PH = SY,FI). Furthermore, since this leaves no indicator variables in the factor loading matrix, Λ_x is specified as zero (LX = ZE). Finally, the error variance/covariance matrix is specified as a diagonal matrix (TD = DI).

TABLE 5.5. LISREL Input and Parameter Specification Summary for Model 1 (Null Model)

```
MO NX=12 NK=12 LX=ZE PH=SY,FI TD=DI
ST 1.00 PH(1,1) PH(2,2) PH(3,3) PH(4,4) PH(5,5) PH(6,6) PH(7,7) PH(8,8)
ST 1.00 PH(9,9) PH(10,10) PH(11,11) PH(12,12)
ST .2 TD(1,1) TD(2,2) TD(3,3) TD(4,4) TD(5,5) TD(6,6) TD(7,7) TD(8,8)
ST .2 TD(9,9) TD(10,10) TD(11,11) TD(12,12)
OU NS SE TV MI
```

PARAMETER SPECIFICATIONS

PHI

	SDQGSC	SDQASC	SDQESC	SDQMSC	APIGSC
SDQGSC	0				
SDQASC	0	0			
SDQESC	0	0	0		
SDQMSC	0	0	0	0	
APIGSC	0	0	0	0	0
APIASC	0	0	0	0	0
APIESC	0	0	0	0	0
APIMSC	0	0	0	0	0
SEIGSC	0	0	0	0	0
SCAASC	0	0	0	0	0
SCAESC	0	0	0	0	0
SCAMSC	0	0	0	0	0

	APIASC	APIESC	APIMSC	SEIGSC	SCAASC
	0				
	0	0			
	0	0	0		
	0	0	0	0	
	0	0	0	0	0
	0	0	0	0	0
	0	0	0	0	0

PHI

	SCAESC	SCAMSC
SCAESC	0	
SCAMSC	0	0

THETA DELTA

SDQGSC	SDQASC	SDQESC	SDQMSC	APIGSC
1	2	3	4	5

APIASC	APIESC	APIMSC	SEIGSC	SCAASC
6	7	8	9	10

THETA DELTA

SCAESC	SCAMSC
11	12

TABLE 5.6. LISREL Input and Parameter Specification Summary for Model 2

```
MO NX=12 NK=4 LX=FU PH=SY,FI TD=DI
FR LX(1,1) LX(5,1) LX(9,1) LX(2,2) LX(6,2) LX(10,2) LX(3,3) LX(7,3)
FR LX(11,3) LX(4,4) LX(8,4) LX(12,4)
FR PH(2,1) PH(3,1) PH(4,1) PH(3,2) PH(4,2) PH(4,3)
ST 1.00 PH(1,1) PH(2,2) PH(3,3) PH(4,4)
ST .9 LX(1,1) LX(5,1) LX(9,1) LX(2,2) LX(6,2) LX(10,2) LX(3,3) LX(7,3)
ST .9 LX(11,3) LX(4,4) LX(8,4) LX(12,4)
ST .02 PH(2,1) PH(3,1) PH(4,1) PH(3,2) PH(4,2) PH(4,3)
ST .2 TD(1,1) TD(2,2) TD(3,3) TD(4,4) TD(5,5) TD(6,6) TD(7,7) TD(8,8)
ST .2 TD(9,9) TD(10,10) TD(11,11) TD(12,12)
OU NS SE TV MI
```

PARAMETER SPECIFICATIONS

LAMBDA X

	KSI 1	KSI 2	KSI 3	KSI 4
SDQGSC	1	0	0	0
SDQASC	0	2	0	0
SDQESC	0	0	3	0
SDQMSC	0	0	0	4
APIGSC	5	0	0	0
APIASC	0	6	0	0
APIESC	0	0	7	0
APIMSC	0	0	0	8
SEIGSC	9	0	0	0
SCAASC	0	10	0	0
SCAESC	0	0	11	0
SCAMSC	0	0	0	12

PHI

	KSI 1	KSI 2	KSI 3	KSI 4
KSI 1	0			
KSI 2	13	0		
KSI 3	14	15	0	
KSI 4	16	17	18	0

THETA DELTA

SDQGSC	SDQASC	SDQESC	SDQMSC	APIGSC
19	20	21	22	23

APIASC	APIESC	APIMSC	SEIGSC	SCAASC
24	25	26	27	28

THETA DELTA

SCAESC	SCAMSC
29	30

Model 2 (see Table 5.6) is postulated to have four trait factors that are allowed to correlate; no method factors are specified.[5] As such, the model is specified as having 12 observed variables (NX = 12) that measure four latent factors (NK = 4). The factor variance/covariance matrix (Φ) is specified as symmetric and fixed (PH = SY,FI) with the variances fixed to 1.00 for identification purposes and the covariances left unconstrained [FR PH(2,1)-----PH(4,3)].

---

[5]In fact, the same results are obtained if the model is specified as a seven-factor model, allowing the trait factors to remain fixed at 0.0. This will be illustrated with Model 5.

TABLE 5.7. LISREL Input and Parameter Specification Summary for Model 3

```
MO NX=12 NK=7 LX=FU PH=SY,FI TD=DI
FR LX(1,1) LX(5,1) LX(9,1) LX(2,2) LX(6,2) LX(10,2) LX(3,3) LX(7,3)
FR LX(11,3) LX(4,4) LX(8,4) LX(12,4)
FR LX(1,5) LX(2,5) LX(3,5) LX(4,5) LX(5,5) LX(6,6) LX(7,6) LX(8,6)
FR LX(9,7) LX(10,7) LX(11,7) LX(12,7)
FR PH(2,1) PH(3,1) PH(4,1) PH(3,2) PH(4,2) PH(4,3)
ST 1.00 PH(1,1) PH(2,2) PH(3,3) PH(4,4) PH(5,5) PH(6,6) PH(7,7)
ST .9 LX(1,1) LX(5,1) LX(9,1) LX(2,2) LX(6,2) LX(10,2) LX(3,3) LX(7,3)
ST .9 LX(11,3) LX(4,4) LX(8,4) LX(12,4)
ST .9 LX(1,5) LX(2,5) LX(3,5) LX(4,5) LX(5,6) LX(6,6) LX(7,6) LX(8,6)
ST .9 LX(9,7) LX(10,7) LX(11,7) LX(12,7)
ST .02 PH(2,1) PH(3,1) PH(4,1) PH(3,2) PH(4,2) PH(4,3)
ST .2 TD(1,1) TD(2,2) TD(3,3) TD(4,4) TD(5,5) TD(6,6) TD(7,7) TD(8,8)
ST .2 TD(9,9) TD(10,10) TD(11,11) TD(12,12)
OU NS SE TV MI
```

PARAMETER SPECIFICATIONS

LAMBDA X

	KSI 1	KSI 2	KSI 3	KSI 4	KSI 5
SDQGSC	1	0	0	0	2
SDQASC	0	3	0	0	4
SDQESC	0	0	5	0	6
SDQMSC	0	0	0	7	8
APIGSC	9	0	0	0	0
APIASC	0	11	0	0	0
APIESC	0	0	13	0	0
APIMSC	0	0	0	15	0
SEIGSC	17	0	0	0	0
SCAASC	0	19	0	0	0
SCAESC	0	0	21	0	0
SCAMSC	0	0	0	23	0

KSI 6	KSI 7
0	0
0	0
0	0
0	0
10	0
12	0
14	0
16	0
0	18
0	20
0	22
0	24

PHI

	KSI 1	KSI 2	KSI 3	KSI 4	KSI 5
KSI 1	0				
KSI 2	25	0			
KSI 3	26	27	0		
KSI 4	28	29	30	0	
KSI 5	0	0	0	0	0
KSI 6	0	0	0	0	0
KSI 7	0	0	0	0	0

KSI 6	KSI 7
0	
0	0

TABLE 5.7. Continued

THETA DELTA

$$\underline{SDGGSC}_{31} \qquad \underline{SDJASC}_{32} \qquad \underline{SDGESC}_{33} \qquad \underline{SDGMSC}_{34} \qquad \underline{APIGSC}_{35}$$

$$\underline{APIASC}_{36} \qquad \underline{APIESC}_{37} \qquad \underline{APIMSC}_{38} \qquad \underline{SEIGSC}_{39} \qquad \underline{SCAASC}_{40}$$

THETA DELTA

$$\underline{SCAESC}_{41} \qquad \underline{SCAMSC}_{42}$$

Model 3 (see Table 5.7) is hypothesized as having seven factors (NK = 7): four correlated trait factors and three uncorrelated (i.e., orthogonal) method factors. Thus, while the trait factors are specified as free, the method factors are left fixed at 0.0 by default (PH = SY,FI).

Model 4 (see Table 5.2) is the hypothesized seven-factor model discussed earlier. *Model 5* (see Table 5.8) postulates a three-factor model (NK = 3) with three correlated method factors; no trait factors are specified. Thus in the Phi matrix, the factor covariances are specified as free [PH(2,2), PH(3,2)].

Model 6 (see Table 5.9) postulated a seven-factor model with trait factors perfectly correlated as indicated by the start values for the trait covariances [ST 1.00 PH(2.1)—–-PH(4,3)] and method factors allowed to correlate freely [FR PH(6,5) PH(7,5) PH(7,6)].

In order to determine the extent to which each measurement scale was contributing to method bias, three additional models were postulated for comparison purposes. These are as follows.

In *Model 7* (see Table 5.10), the Likert scale has been deleted. Thus, although it is specified as a seven-factor model, factor loadings for the Likert scale [LX(1,5) LX(2,5) LX(3,5) LX(4,5)] and correlations between the Likert and the other scales [PH(6,5) PH(7,5)] have been fixed to 0.0. Here again, this model, as well as Models 8 and 9 could have been alterna-

TABLE 5.8. LISREL Input and Parameter Specification Summary for Model 5

```
MO NX=12 NK=7 LX=FU PH=SY,FI TD=DI
FR LX(1,5) LX(2,5) LX(3,5) LX(4,5) LX(5,5) LX(6,6) LX(7,6) LX(8,6)
FR LX(9,7) LX(10,7) LX(11,7) LX(12,7)
FR PH(6,5) PH(7,5) PH(7,6)
ST 1.00 PH(1,1) PH(2,2) PH(3,3) PH(4,4) PH(5,5) PH(6,6) PH(7,7)
ST .3 LX(1,5) LX(2,5) LX(3,5) LX(4,5) LX(5,6) LX(6,6) LX(7,6) LX(8,6)
ST .3 LX(9,7) LX(10,7) LX(11,7) LX(12,7)
ST .02 PH(6,5) PH(7,5) PH(7,6)
ST .2 TD(1,1) TD(2,2) TD(3,3) TD(4,4) TD(5,5) TD(6,6) TD(7,7) TD(8,8)
ST .2 TD(9,9) TD(10,10) TD(11,11) TD(12,12)
OU NS SE TV MI
```

TABLE 5.8. Continued

PARAMETER SPECIFICATIONS

LAMBDA X

	KSI 1	KSI 2	KSI 3	KSI 4	KSI 5
SDQGSC	0	0	0	0	1
SDQASC	0	0	0	0	2
SDQESC	0	0	0	0	3
SDQMSC	0	0	0	0	4
APIGSC	0	0	0	0	0
APIASC	0	0	0	0	0
APIESC	0	0	0	0	0
APIMSC	0	0	0	0	0
SEIGSC	0	0	0	0	0
SCAASC	0	0	0	0	0
SCAESC	0	0	0	0	0
SCAMSC	0	0	0	0	0

	KSI 6	KSI 7
	0	0
	0	0
	0	0
	0	0
	5	0
	6	0
	7	0
	8	0
	0	9
	0	10
	0	11
	0	12

PHI

	KSI 1	KSI 2	KSI 3	KSI 4	KSI 5
KSI 1	0				
KSI 2	0	0			
KSI 3	0	0	0		
KSI 4	0	0	0	0	
KSI 5	0	0	0	0	0
KSI 6	0	0	0	0	13
KSI 7	0	0	0	0	14

	KSI 6	KSI 7
	0	
	15	0

THETA DELTA

$SDQGSC_{16}$	$SDQASC_{17}$	$SDQESC_{18}$	$SDQMSC_{19}$	$APIGSC_{20}$
$APIASC_{21}$	$APIESC_{22}$	$APIMSC_{23}$	$SEIGSC_{24}$	$SCAASC_{25}$

THETA DELTA

$SCAESC_{26}$	$SCAMSC_{27}$

TABLE 5.9. LISREL Input and Parameter Specification Summary for Model 6

```
MO NX=12 NK=7 LX=FU PH=SY.FI TD=DI
FR LX(1,1) LX(5,1) LX(9,1) LX(2,2) LX(6,2) LX(10,2) LX(3,3) LX(7,3)
FR LX(11,3) LX(4,4) LX(8,4) LX(12,4)
FR LX(1,5) LX(2,5) LX(3,5) LX(4,5) LX(5,5) LX(6,6) LX(7,6) LX(8,6)
FR LX(9,7) LX(10,7) LX(11,7) LX(12,7)
FR PH(6,5) PH(7,5) PH(7,6)
ST 1.00 PH(1,1) PH(2,2) PH(3,3) PH(4,4) PH(5,5) PH(6,6) PH(7,7)
ST .9 LX(1,1) LX(5,1) LX(9,1) LX(3,3) LX(7,3)
ST -.4 LX(2,2) LX(6,2) LX(10,2)
ST .9 LX(11,3) LX(4,4) LX(8,4) LX(12,4)
ST .3 LX(1,5) LX(2,5) LX(3,5) LX(4,5) LX(5,6) LX(6,6) LX(7,6) LX(8,6)
ST .3 LX(9,7) LX(10,7) LX(11,7) LX(12,7)
ST 1.0 PH(2,1) PH(3,1) PH(4,1) PH(3,2) PH(4,2) PH(4,3)
ST .02 PH(6,5) PH(7,5) PH(7,6)
ST .2 TD(1,1) TD(2,2) TD(3,3) TD(4,4) TD(5,5) TD(6,6) TD(7,7) TD(8,8)
ST .2 TD(9,9) TD(10,10) TD(11,11) TD(12,12)
OU NS SE TV MI
```

PARAMETER SPECIFICATIONS

LAMBDA X

	KSI 1	KSI 2	KSI 3	KSI 4	KSI 5
SDQGSC	1	0	0	0	2
SDQASC	0	3	0	0	4
SDQESC	0	0	5	0	6
SDQMSC	0	0	0	7	8
APIGSC	9	0	0	0	0
APIASC	0	11	0	0	0
APIESC	0	0	13	0	0
APIMSC	0	0	0	15	0
SEIGSC	17	0	0	0	0
SCAASC	0	19	0	0	0
SCAESC	0	0	21	0	0
SCAMSC	0	0	0	23	0

KSI 6	KSI 7
0	0
0	0
0	0
0	0
10	0
12	0
14	0
16	0
0	18
0	20
0	22
0	24

PHI

	KSI 1	KSI 2	KSI 3	KSI 4	KSI 5
KSI 1	0				
KSI 2	0	0			
KSI 3	0	0	0		
KSI 4	0	0	0	0	
KSI 5	0	0	0	0	0
KSI 6	0	0	0	0	25
KSI 7	0	0	0	0	26

KSI 6	KSI 7
0	
27	0

TABLE 5.9. Continued

THETA DELTA

$$\underline{SDQGSC}_{28} \qquad \underline{SDQASC}_{29} \qquad \underline{SDQESC}_{30} \qquad \underline{SDQMSC}_{31} \qquad \underline{APIGSC}_{32}$$

$$\underline{APIASC}_{33} \qquad \underline{APIESC}_{34} \qquad \underline{APIMSC}_{35} \qquad \underline{SEIGSC}_{36} \qquad \underline{SCAASC}_{37}$$

THETA DELTA

$$\underline{SCAESC}_{38} \qquad \underline{SCAMSC}_{39}$$

TABLE 5.10. LISREL Input and Parameter Specification Summary for Model 7

```
MO NX=12 NK=7 LX=FU PH=SY,FI TD=DI
FR LX(1,1) LX(5,1) LX(9,1) LX(2,2) LX(6,2) LX(10,2) LX(3,3) LX(7,3)
FR LX(11,3) LX(4,4) LX(8,4) LX(12,4)
FR LX(5,6) LX(6,6) LX(7,6) LX(8,6)
FR LX(9,7) LX(10,7) LX(11,7) LX(12,7)
FR PH(2,1) PH(3,1) PH(4,1) PH(3,2) PH(4,2) PH(4,3)
FR PH(7,6)
FI TD(5,5)
ST 1.00 PH(1,1) PH(2,2) PH(3,3) PH(4,4) PH(5,5) PH(6,6) PH(7,7)
ST .9 LX(1,1) LX(5,1) LX(9,1) LX(3,3) LX(7,3)
ST -.4 LX(2,2) LX(6,2) LX(10,2)
ST .9 LX(11,3) LX(4,4) LX(8,4) LX(12,4)
ST .9 LX(5,6) LX(6,6) LX(7,6) LX(8,6)
ST .9 LX(9,7) LX(10,7) LX(11,7) LX(12,7)
ST .02 PH(2,1) PH(3,1) PH(4,1) PH(3,2) PH(4,2) PH(4,3)
ST .02 PH(7,6)
ST .2 TD(1,1) TD(2,2) TD(3,3) TD(4,4) TD(6,6) TD(7,7) TD(8,8)
ST .2 TD(9,9) TD(10,10) TD(11,11) TD(12,12)
ST .01 TD(5,5)
OU NS SE TV MI
```

PARAMETER SPECIFICATIONS

LAMBDA X

	KSI 1	KSI 2	KSI 3	KSI 4	KSI 5
SDQGSC	1	0	0	0	0
SDQASC	0	2	0	0	0
SDQESC	0	0	3	0	0
SDQMSC	0	0	0	4	0
APIGSC	5	0	0	0	0
APIASC	0	7	0	0	0
APIESC	0	0	9	0	0
APIMSC	0	0	0	11	0
SEIGSC	13	0	0	0	0
SCAASC	0	15	0	0	0
SCAESC	0	0	17	0	0
SCAMSC	0	0	0	19	0

	KSI 6	KSI 7
	0	0
	0	0
	0	0
	0	0
	6	0
	8	0
	10	0
	12	0
	0	14
	0	16
	0	18
	0	20

TABLE 5.10. Continued

PHI

	KSI_1	KSI_2	KSI_3	KSI_4	KSI_5
KSI 1	0				
KSI 2	21	0			
KSI 3	22	23	0		
KSI 4	24	25	26	0	
KSI 5	0	0	0	0	0
KSI 6	0	0	0	0	0
KSI 7	0	0	0	0	0

	KSI_6	KSI_7
KSI 6	0	
KSI 7	27	0

THETA DELTA

$SDQGSC_{28}$	$SDQASC_{29}$	$SDQESC_{30}$	$SDQMSC_{31}$	$APIGSC_{0}$
$APIASC_{32}$	$APIESC_{33}$	$APIMSC_{34}$	$SEIGSC_{35}$	$SCAASC_{36}$

THETA DELTA

$SCAESC_{37}$	$SCAMSC_{38}$

tively specified as six-factor models; as such, it would not be necessary to constrain the deleted scale parameters to 0.0.

In *Model 8* (see Table 5.11), the semantic differential scale has been deleted. Accordingly, the related parameters have been fixed to 0.0 [LX(5,6) LX(6,6) LX(7,6) LX8,6) PH(6,5) PH(7,6)].

Finally, in *Model 9* (see Table 5.12), the Guttman scale has been eliminated. As such, the following parameters were fixed to 0.0: LX(9,7) LX(10,7) LX(11,7) LX(12,7) PH(7,5) PH(7,6). Note that here, the φs are fixed to 0.0 by default.

Goodness-of-fit indices for each of these MTMM models are summarized in Table 5.13. As shown here, Model 1, the most restrictive model, serves as a null model against which competing models are compared in order to determine the BBI. As expected, the hypothesized seven-factor model (Model 4) exhibited the best fit to the data.

4.2. Results of MTMM Model Comparisons

All model comparisons, together with their goodness-of-fit indices are presented in Table 5.14.

TABLE 5.11. LISREL Input and Parameter Specification Summary for Model 8

```
MO  NX=12 NK=7 LX=FU PH=SY,FI TD=DI
FR  LX(1,1) LX(5,1) LX(9,1) LX(2,2) LX(6,2) LX(10,2) LX(3,3) LX(7,3)
FR  LX(11,3) LX(4,4) LX(8,4) LX(12,4)
FR  LX(1,5) LX(2,5) LX(3,5) LX(4,5)
FR  LX(9,7) LX(10,7) LX(11,7) LX(12,7)
FR  PH(2,1) PH(3,1) PH(4,1) PH(3,2) PH(4,2) PH(4,3)
FR  PH(7,5)
FI  TD(5,5)
ST  1.00 PH(1,1) PH(2,2) PH(3,3) PH(4,4) PH(5,5) PH(6,6) PH(7,7)
ST  .9 LX(1,1) LX(5,1) LX(9,1) LX(3,3) LX(7,3)
ST  -.4 LX(2,2) LX(6,2) LX(10,2)
ST  .9 LX(11,3) LX(4,4) LX(8,4) LX(12,4)
ST  .9 LX(1,5) LX(2,5) LX(3,5) LX(4,5)
ST  .9 LX(9,7) LX(10,7) LX(11,7) LX(12,7)
ST  .02 PH(2,1) PH(3,1) PH(4,1) PH(3,2) PH(4,2) PH(4,3)
ST  .02 PH(7,5)
ST  .2 TD(1,1) TD(2,2) TD(3,3) TD(4,4) TD(6,6) TD(7,7) TD(8,8)
ST  .2 TD(9,9) TD(10,10) TD(11,11) TD(12,12)
ST  .01 TD(5,5)
OU  NS SE TV MI
```

PARAMETER SPECIFICATIONS

   LAMBDA X

	KSI 1	KSI 2	KSI 3	KSI 4	KSI 5
SDQGSC	1	0	0	0	2
SDQASC	0	3	0	0	4
SDQESC	0	0	5	0	6
SDQMSC	0	0	0	7	8
APIGSC	9	0	0	0	0
APIASC	0	10	0	0	0
APIESC	0	0	11	0	0
APIMSC	0	0	0	12	0
SEIGSC	13	0	0	0	0
SCAASC	0	15	0	0	0
SCAESC	0	0	17	0	0
SCAMSC	0	0	0	19	0

	KSI 6	KSI 7
	0	0
	0	0
	0	0
	0	0
	0	0
	0	0
	0	0
	0	0
	0	14
	0	16
	0	18
	0	20

   PHI

	KSI 1	KSI 2	KSI 3	KSI 4	KSI 5
KSI 1	0				
KSI 2	21	0			
KSI 3	22	23	0		
KSI 4	24	25	25	0	
KSI 5	0	0	0	0	0
KSI 6	0	0	0	0	0
KSI 7	0	0	0	0	27

	KSI 6	KSI 7
	0	
	0	0

TABLE 5.11. Continued

THETA DELTA

$\underline{SDQGSC}_{28}$ $\underline{SDQASC}_{29}$ $\underline{SDQESC}_{30}$ $\underline{SDQMSC}_{31}$ $\underline{APIGSC}_{0}$

$\underline{APIASC}_{32}$ $\underline{APIESC}_{33}$ $\underline{APIMSC}_{34}$ $\underline{SEIGSC}_{35}$ $\underline{SCAASC}_{36}$

THETA DELTA

$\underline{SCAESC}_{37}$ $\underline{SCAMSC}_{38}$

TABLE 5.12. LISREL Input and Parameter Specification Summary for Model 9

```
MO NX=12 NK=7 LX=FU PH=SY,FI TD=DI
FR LX(1,1) LX(5,1) LX(2,2) LX(9,1) LX(6,2) LX(10,2) LX(3,3) LX(7,3)
FR LX(11,3) LX(4,4) LX(8,4) LX(12,4)
FR LX(1,5) LX(2,5) LX(3,5) LX(4,5) LX(5,5) LX(6,6) LX(7,6) LX(8,6)
FR PH(2,1) PH(3,1) PH(4,1) PH(3,2) PH(4,2) PH(4,3)
FR PH(6,5)
FI TD(5,5)
ST 1.00 PH(1,1) PH(2,2) PH(3,3) PH(4,4) PH(5,5) PH(6,6) PH(7,7)
ST .9 LX(1,1) LX(5,1) LX(9,1) LX(3,3) LX(7,3)
ST -.4 LX(2,2) LX(6,2) LX(10,2)
ST .9 LX(11,3) LX(4,4) LX(8,4) LX(12,4)
ST .9 LX(2,5) LX(5,6) LX(6,6) LX(7,6) LX(8,6)
ST .9 LX(1,5) LX(3,5) LX(4,5)
ST .02 PH(2,1) PH(3,1) PH(3,2) PH(4,1) PH(4,2) PH(4,3)
ST .02 PH(6,5)
ST .2 TD(1,1) TD(2,2) TD(3,3) TD(4,4) TD(6,6) TD(7,7) TD(8,8)
ST .2 TD(9,9) TD(10,10) TD(11,11) TD(12,12)
ST .01  TD(5,5)
OU NS SE TV MI
```

PARAMETER SPECIFICATIONS

LAMBDA X

	KSI 1	KSI 2	KSI 3	KSI 4	KSI 5
SDQGSC	1	0	0	0	2
SDQASC	0	3	0	0	4
SDQESC	0	0	5	0	6
SDQMSC	0	0	0	7	8
APIGSC	9	0	0	0	0
APIASC	0	11	0	0	0
APIESC	0	0	13	0	0
APIMSC	0	0	0	15	0
SEIGSC	17	0	0	0	0
SCAASC	0	18	0	0	0
SCAESC	0	0	19	0	0
SCAMSC	0	0	0	20	0

KSI 6	KSI 7
0	0
0	0
0	0
0	0
10	0
12	0
14	0
16	0
0	0
0	0
0	0
0	0

TABLE 5.12. Continued

PHI

	KSI 1	KSI 2	KSI 3	KSI 4	KSI 5
KSI 1	0				
KSI 2	21	0			
KSI 3	22	23	0		
KSI 4	24	25	26	0	
KSI 5	0	0	0	0	0
KSI 6	0	0	0	0	27
KSI 7	0	0	0	0	0

	KSI 6	KSI 7
	0	
	0	0

THETA DELTA

$SDGGSC_{28}$	$SDJASC_{29}$	$SDGESC_{30}$	$SDGMSC_{31}$	$APIGSC_{0}$
$APIASC_{32}$	$APIESC_{33}$	$APIMSC_{34}$	$SEIGSC_{35}$	$SCAASC_{36}$

THETA DELTA

$SCAESC_{37}$	$SCAMSC_{38}$

Let's begin with a comparison of models to determine the strength of added components in the hypothesized model. For example, by comparing Model 3 with Model 4, we can determine the impact of correlations among the method factors (as with comparisons between Models 2 and 3). By so doing, we see that although Model 4 hypothesizes correlations among both the trait and method factors, method correlations appear to be relatively weak ($\Delta\chi^2_{(3)}$ = 9.48; p <0.05; $\Delta\chi^2$/df = 0.0; ΔBBI = 0.02). These results suggest minimal method-related variance in that the three measurement scales are operating independently and support findings from our earlier investigation of the individual method parameters.

To test for convergent validity, we now compare Model 4 with Model 5 in which no trait factors are specified. A significant difference between these two models argues for the presence of trait factors, and thus for evidence of convergent validity. As shown in Table 5.14, the $\Delta\chi^2$ was highly significant, thus providing strong evidence of convergent validity for the trait factors.

Since the discriminant validity of traits argues for their zero intercorrelations, evidence of same can be tested by comparing the baseline model (Model 4) with one in which perfect correlations among traits are hypothesized (Model 6). The highly significant difference resulting from this

TABLE 5.13. Goodness-of-Fit Indices for Multitrait-Multimethod Models

Model	χ^2	df	χ^2/df	BBI
1. 12 uncorrelated factors (null model)	1,681.05	66	25.47	—
2. 4 correlated trait factors no method factors	216.26	48	4.51	.871
3. 4 correlated trait factors 3 uncorrelated method factors	114.69	36	3.19	.914
4. 4 correlated trait factors 3 correlated method factors (baseline model)	105.21	33	3.19	.937
5. no trait factors 3 correlated method factors	868.09	51	17.02	.484
6. 4 perfectly correlated trait factors, freely correlated method factors	403.61	39	10.35	.760
7. 4 correlated trait factors 2 correlated method factors (semantic differential, Guttman)	154.14	39	3.95	.908
8. 4 correlated trait factors 2 correlated method factors (Likert, Guttman)	110.73	39	2.83	.932
9. 4 correlated trait factors 2 correlated method factors (Likert, semantic differential)	133.00	39	3.41	.921

From Byrne (in press), "Multigroup Comparisons and the Assumption of Equivalent Construct Validity Across Groups: Methodological and Substantive Issues" in *Multivariate Behavioral Research*. Copyright 1989 by Lawrence Erlbaum Associates, Inc. Reprinted with permission.

comparison argues for strong evidence of discriminant validity for the traits in the hypothesized model. The discriminant validity of method factors (i.e., no method bias) can be tested by comparing Model 4 with Model 2 in which no method factors are specified. While this comparison yielded a statistically significant $\Delta\chi^2$, suggesting evidence of method bias, this effect was substantially weaker than that related to the trait factors.

Finally, to determine the extent to which each measurement scale is contributing to the method bias, we can compare Model 4 with Models 7, 8, and 9 in which the Likert, semantic differential, and Guttman scales, respectively, have been deleted. These results show significant method effects for the Likert and Guttman scales; those associated with the semantic differential were not significant. Overall, the Likert scale appears to be the heaviest contributor to method bias. It must be pointed out, however, that these three model comparisons were conducted for illustrative purposes only. In actual fact, differences in the subjective fit indices demonstrated negligible method effects.

TABLE 5.14. Goodness-of-Fit Indices for Comparison of Multitrait-Multimethod Models[a]

Model comparison	Differences in			
	χ^2	df	χ^2/df	BBI
Tests of Added Components				
Model 1 vs Model 2	1,464.79	18	20.96	—
Model 2 vs Model 3	101.57	12	.96	.04
Model 3 vs Model 4	9.48[*]	3	0.00	.02
Test of Convergent Validity				
Model 4 vs Model 5 (traits)	762.88	18	13.83	.45
Tests of Discriminant Validity				
Model 4 vs Model 6 (traits)	298.40	6	7.16	.18
Model 4 vs Model 2 (methods)	111.05	15	1.32	.07
Tests of Method Bias				
Model 4 vs Model 7 (Likert)	48.93	6	.76	.03
Model 4 vs Model 8 (semantic differential)	5.52[b]	6	.36	.00
Model 4 vs Model 9 (Guttman)	27.79	6	.22	.02

[*]$p < 0.05$
[a]Unasterisked χ^a difference values are statistically significant at $p < 0.001$.
[b]Not statistically significant.
From Byrne (in press), "Multigroup Comparisons and the Assumption of Equivalent Construct Validity Across Groups: Methodological and Substantive Issues" in *Multivariate Behavioral Research*. Copyright 1989 by Lawrence Erlbaum Associates, Inc. Reprinted with permission.

5. Summary

This chapter focused on the analysis of a multitrait-multimethod matrix using the LISREL methodology in determining evidence of convergent and discriminant validity. Two possible means to deriving this evaluative information were demonstrated: the examination of individual parameters related to trait and methods factors and the comparison of alternatively specified models in which various aspects of construct validity are considered. Finally, the comparison of competing models was used to identify the extent of method bias associated with particular measuring instruments.

Section III   Multigroup Analyses

6
Testing for Measurement and Structural Invariance of a Theoretical Construct

In this section, we focus on the analysis of CFA models comprising two groups of subjects. Of primary interest are procedures involved in testing for the invariance (i.e., equivalence) of measurements and/or structure across two or more groups.

In our first multigroup application, we test hypotheses related to equivalencies across gender related to multiple SC measurements and the factorial structure of a multidimensional SC comprising general SC, academic SC, English SC, and mathematics SC. This work follows from the previous study examined in Application 1, Section II. (For details of the study related to this application, see Byrne & Shavelson, 1987.)

1. Testing for Factorial Invariance: The General Framework

1.1. Preliminary Single-Group Analyses

As a prerequisite to testing for factorial invariance, it is convenient to consider a baseline model that is estimated separately for each group. As noted in Section II, this model represents the most parsimonious, yet substantively most meaningful and best-fitting model to the data. Since the goodness-of-fit value and its corresponding degrees of freedom are additive, the sum of the χ^2s reflects how well the underlying factor structure fits the data across groups.

However, since instruments are often group-specific in the way they operate, baseline models are not expected to be identical across groups. For example, whereas the baseline model for one group might include correlated measurement errors and/or secondary factor loadings, this may not be so for a second group. A priori knowledge of such group differences, as will be illustrated later, is critical to the conduct of invariance testing procedures. Although the bulk of the literature suggests that the number of factors must be equivalent across groups before further tests of invariance can be conducted, this is only a logical starting point, not a necessary condition; only the comparable parameters within the same factor need be equated (Werts, Rock, Linn, & Jöreskog, 1976).

Since the estimation of baseline models involves no between-group constraints, the data may be analyzed separately for each group. In testing for invariance, however, constraints are imposed on particular parameters, and thus the data from all groups must be analyzed simultaneously to obtain efficient estimates (Jöreskog & Sorböm, 1985); the pattern of fixed and free parameters remaining consistent with that specified in the baseline model for each group.

Tests of factorial invariance, then, can involve both measurement and structural components of a model. In LISREL notation, this means that the factor (lambda, Λ), error (theta, Θ), and latent factor variance-covariance (phi, Φ) matrices are of primary interest. If, however, the invariance testing includes factor means, then the regression intercept (nu ν) and mean (gamma, Γ) vectors are also of interest; this issue is addressed in Application 6 (see Chapter 8).

1.2. Subsequent Multigroup Analyses

Tests of factorial invariance can begin with an overall test of the equality of covariance structures across groups (i.e., H_0: $\Sigma_1 = \Sigma_2 = \ldots \Sigma_G$ where G is the number of groups). As such, failure to reject the null hypothesis is interpreted as evidence of invariance across groups; except for mean structures, the groups can be treated as one. This means, then, that the variance-covariance matrices can be pooled and that subsequent investigative inquiry would be based on single-group analyses; there is no need to analyze each group separately or simultaneously. Rejection of this hypothesis, on the other hand, argues for the testing of a series of increasingly restrictive hypotheses in order to identify the source of nonequivalence.

Unfortunately, the global test of invariant variance-covariance matrices, while seemingly straightforward and reasonable, often leads to contradictory findings thereby contributing more to confusion than to enlightenment with respect to equivalencies across groups. For example, sometimes this initial hypothesis is found tenable, yet subsequent tests for the invariance of particular measurement and/or structural parameters must be rejected (see e.g., Jöreskog, 1971a). Conversely, this initial hypothesis may be rejected, yet the measurement and/or measurement parameters may be found to be group-invariant. Thus, while Jöreskog suggested that the global hypothesis of equivalent covariance matrices be tested first, before proceeding to more specific hypotheses, he is nonetheless cognizant of various problems associated with its application. Furthermore, Rock, Werts, and Flaugher (1978) have advocated that even if this hypothesis cannot be rejected, the researcher should still conduct subsequent tests for the invariance of particular parameters.

Such inconsistencies in the omnibus test of invariance derive from the fact that there is no baseline model for the test of invariant variance-co-variance matrices. As a result, this test becomes much more stringent than is the case with subsequent tests of invariance related to the factors (Muthén, personal communication, Oct. 1988). Consequently, Muthén contends that in general the omnibus test is of little import or assistance in testing for invariance across groups, and thus is not a necessary prereq-uisite to the conduct of relatedly more specific hypotheses bearing on factorial invariance.

Let us now continue, then, with tests of hypotheses related to factorial invariance. Specifically, these hypotheses focus on the invariance of: (1) the number of factors (i.e., H_o: $\Lambda_{1K} = \Lambda_{2K} = \ldots \Lambda_{GK}$, where k = number of factors); (2) the factor-loading pattern (i.e., H_o: $\Lambda_1 = \Lambda_2 = \ldots \Lambda_G$); (3) the factor variances and covariances (i.e., H_o: $\Phi_1 = \Phi_2 = \ldots \Phi_G$); and (4) the error/uniquenesses (i.e., H_o: $\Theta_1 = \Theta_2 = \ldots \Theta_G$). The tenabil-ity of Hypotheses 1 and 2 is a logical prerequisite to the testing of Hypoth-eses 3 and 4.

The procedures for testing the invariance hypotheses are identical to those used in model fitting. That is, a model in which certain parameters are constrained to be equal across groups is compared with a less restric-tive model in which these same parameters are free to take on any value. For example, the hypothesis of an invariant pattern of factor loadings (Λ) can be tested by constraining all corresponding lambda parameters to be equal across groups, and then comparing this model with one in which the number of factors and pattern of loadings are invariant, but not con-strained equal (i.e., the summed χ^2 across groups). If the difference in χ^2 ($\Delta\chi^2$) is not significant, the hypothesis of an invariant pattern of loadings is considered tenable.

2. Tests for Invariance Related to Self-Concept

In our first two-group application, then, we begin by establishing a well-fitting baseline model separately for males and females. But first, let's examine the model to be tested.

2.1. The Hypothesized Model

The hypothesized model in the present application is identical to the one examined in Chapter 3 (see Tables 3.2 and 3.3); as such, the API has been deleted as one measure of ASC. To recapitulate, this model hypothesizes a priori that: SC responses can be explained by four factors (general SC, academic SC, English SC, and mathematics SC); each subscale measure has a nonzero loading on the SC factor that it is designed to measure (i.e., target loading) and a zero loading on all other factors (i.e., nontarget loadings); the four SC factors, consistent with the theory (see e.g., Byrne

& Shavelson, 1986), are correlated; and error/uniqueness terms for each of the measures are uncorrelated. To refresh our memory of the model to be tested here, let's review Table 3.3 which summarizes the pattern of parameters to be estimated. Recall that the λs, ϕs, and θ_δs represent the parameters to be estimated and the 0s and 1s are fixed a priori; and for purposes of identification, the first of each congeneric set of SC measures is fixed to 1.0, each nontarget loading is fixed to 0.0. A schematic presentation of the model to be tested is shown in Figure 6.1.

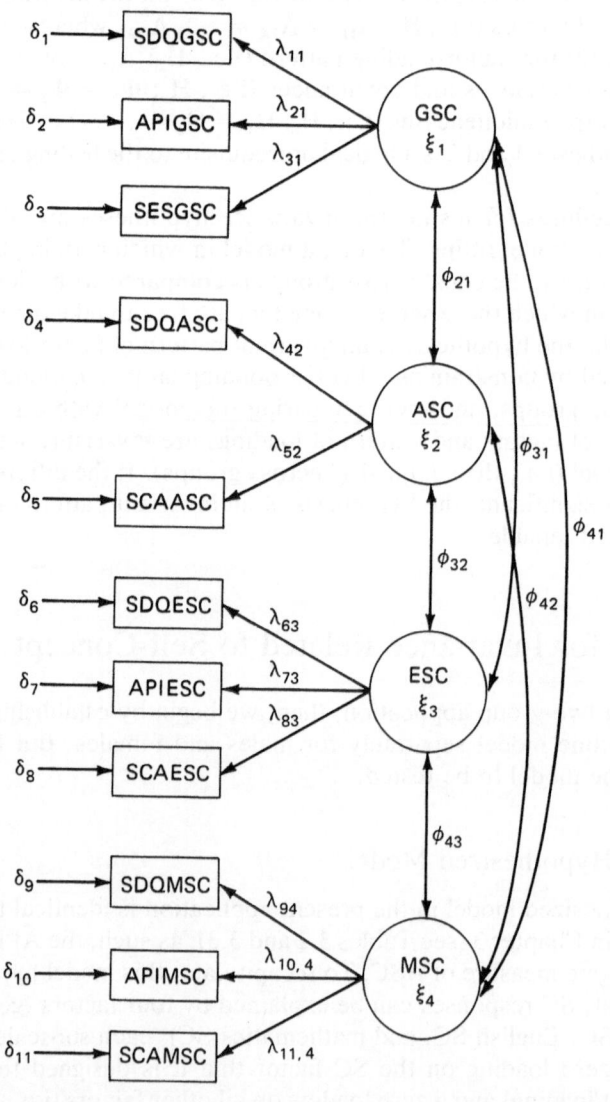

FIGURE 6.1. Hypothesized Four-Factor Model of Self-Concept.

2.2. The Baseline Models

Using the procedures outlined in Chapter 3, alternative models were sequentially respecified and reestimated until a well-fitting model, both statistically and substantively, was found for each sex. These baseline models demonstrate a satisfactory fit to the data for both males (χ^2/df = 1.97; BBI = 0.98) and females (χ^2/df = 2.27; BBI = 0.98).[1]

In fitting the baseline model for each sex, a substantial drop in χ^2 was found when the English SC subscale of the SDQ was free to load on general SC. Furthermore, for males only, the mathematics SC subscale of the SDQ was allowed to load on the English SC factor.[2] Finally, error/ uniquenesses between subscales of the same measuring instrument were allowed to covary, resulting in five error covariances for males and three for females. These baseline models are shown schematically in Figure 6.2.

3. LISREL Input for Multigroup Analyses

As noted earlier, when the analyses focus on multigroup comparisons with constraints between the groups (i.e., certain parameters are constrained equal), it is imperative that the parameters be estimated simultaneously.[3] To this end, in the present application, we now combine our male and female files into one, such that the specifications for each group are stacked one after the other (i.e., the model specifications for males are presented first, followed by the model specifications for females) and these specifications are consistent with the established baseline model for each sex (see Figure 6.2).

3.1. The DA Card

A multiple group specification requires three modifications to the basic setup as it relates to single-group analyses; these are as follows.

---

[1]As noted in Byrne & Shavelson (1987), several alternative models were estimated beyond this point yielding statistically better fitting models for both males (χ^2/df = 1.33; BBI = 0.99) and females (χ^2/df = 1.58; BBI = 0.99). These models, however, allowed for error covariances between subscales of different measuring instruments. Clearly, such covariation does not make sense psychometrically. These models were thus rejected in favor of the more substantively sound baseline models presented here.

[2]From a substantive point of view, this factor loading is difficult to explain. However, as noted in our summary of the study, we strongly suspect that this effect will disappear upon replication.

[3]KM was specified in estimating model parameters for males and females. Thus, the KM specification was changed to CM in testing for invariance across sex. In order for the analyses to be based on the covariance matrix, the standard deviations must be added below the correlation data matrix.

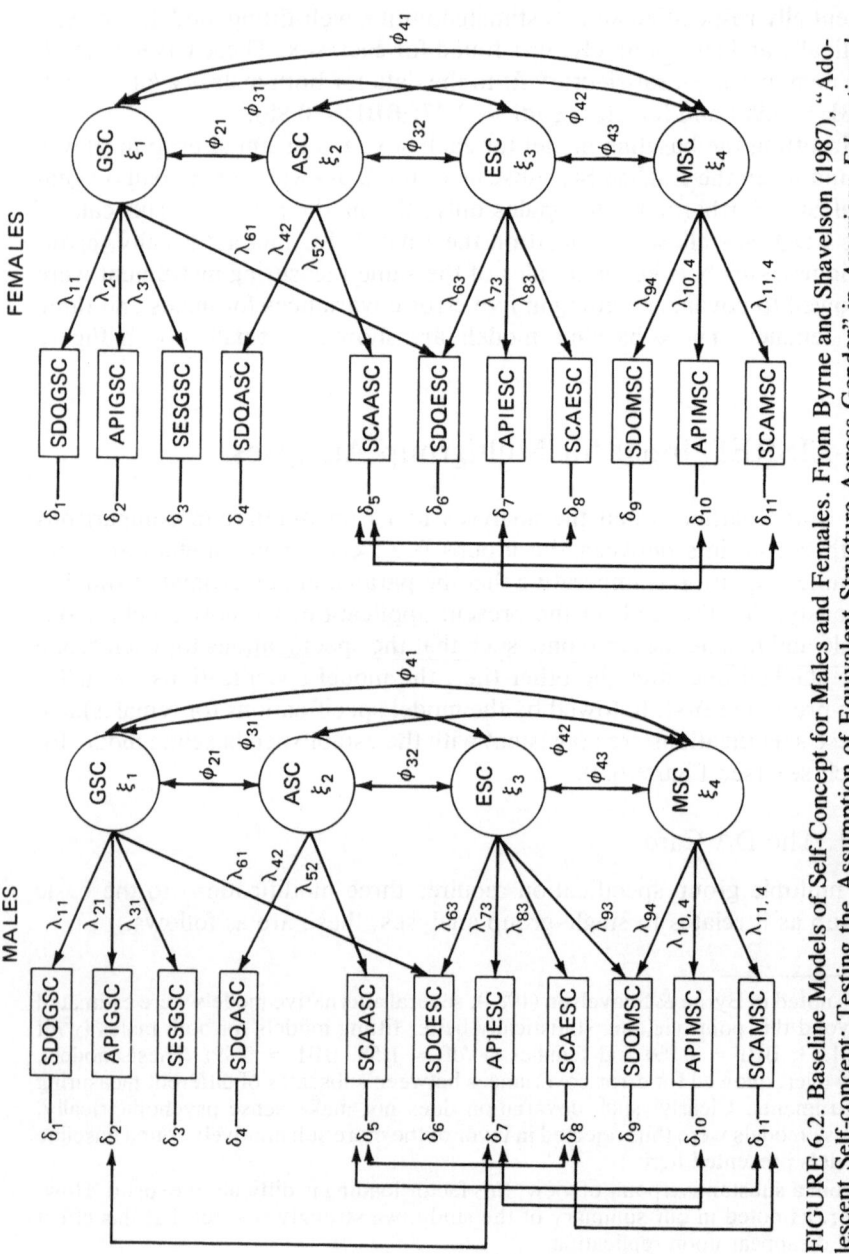

FIGURE 6.2. Baseline Models of Self-Concept for Males and Females. From Byrne and Shavelson (1987), "Adolescent Self-concept: Testing the Assumption of Equivalent Structure Across Gender" in *American Educational Research Journal, 24*(3), 365–385. Copyright 1987 by American Educational Research Association. Reprinted with permission.

1. The number of groups (NG) must be specified for the first group. In the present case, it would be specified as NG = 2.
2. The matrix to be analyzed must be the covariance matrix (CM), rather than the correlation matrix (KM) as demonstrated for the single-group analyses (see e.g., Table 3.2); this specification is made for the first group only and reads MA = CM.
3. The only information required on the DA card of the second group relates to the number of observations expected (N = 420).

These stacked specifications for males and females, based on their baseline models, are presented in Table 6.1.

TABLE 6.1. LISREL Input for the Simultaneous Estimation of Male and Female Baseline Models

```
TESTING INVARIANCE - GRP1=MALES - "LISRELMF" FILE
DA NG=2 NI=15 NO=412 MA=CM
LA
*
'SDQGSC' 'SDQASC' 'SDQESC' 'SDQMSC' 'APIGSC' 'SESGSC' 'APIASC' 'SCAASC'
'APIESC' 'SCAESC' 'APIMSC' 'SCAMSC' 'GPA' 'ENG' 'MATH'
KM SY
(15F4.3)
1000
 3451000
 287 3771000
 252 579 1141000
 637 335 253 2551000
 768 339 302 278 6001000
 543 640 382 451 656 5661000
 254 703 337 591 312 313 5871000
 166 429 724 143 264 216 445 3641000
 104 411 506 142 199 163 373 551 63010000
 247 601 202 880 288 309 530 621 290 1991000
 208 526 127 827 253 231 473 676 155 234 8081000
 035 517 117 481 067 080 374 687 121 340 434 5241000
-020 410 158 272 003 062 297 507 215 492 261 294 7861000
 040 378 060 609 076 074 295 525 077 172 560 649 746 5331000
ME
*
76.410 52.978 55.621 49.223 78.223 32.022 71.175 28.359 57.918 26.820
47.398 26.243 65.546 62.522 59.922
SD
*
13.856 13.383 10.017 15.977 9.442 4.889 9.999 5.943 11.099 5.929 11.762
7.873 11.232 12.664 16.898
SELECTION
1 5 6 2 8 3 9 10 4 11 12/
MO NX=11 NK=4 LX=FU PH=SY TD=SY,FI
FR LX(2,1) LX(3,1) LX(5,2) LX(7,3) LX(8,3) LX(10,4) LX(11,4)
FR LX(6,1) LX(9,3)
FR TD(1,1) TD(2,2) TD(3,3) TD(4,4) TD(5,5) TD(6,6) TD(7,7) TD(8,8)
FR TD(9,9) TD(10,10) TD(11,11)
FR TD(8,5) TD(10,7) TD(11,5) TD(11,8) TD(7,2)
ST 30.0 LX(1,1) LX(4,2) LX(6,3) LX(9,4)
ST 15.0 LX(2,1) LX(3,1) LX(5,2) LX(7,3) LX(8,3) LX(10,4) LX(11,4)
ST 5.0 LX(6,1)
ST -5.0 LX(9,3)
ST .1 PH(1,1) PH(2,2) PH(3,3) PH(4,4)
ST .05 PH(2,1) PH(3,1) PH(4,1) PH(3,2) PH(4,2) PH(4,3)
ST 40.0 TD(1,1) TD(2,2) TD(4,4) TD(6,6)
ST 15.0 TD(3,3) TD(5,5) TD(7,7) TD(8,8) TD(9,9) TD(10,10) TD(11,11)
ST 6.0 TD(8,5) TD(10,7) TD(11,5) TD(11,8) TD(7,2)
OU NS
TESTING EQUALITY - FEMALES
DA NO=420
LA
*
'SDQGSC' 'SDQASC' 'SDQESC' 'SDQMSC' 'APIGSC' 'SESGSC' 'APIASC' 'SCAASC'
'APIESC' 'SCAESC' 'APIMSC' 'SCAMSC' 'GPA' 'ENG' 'MATH'
```

TABLE 6.1. Continued

```
KM SY
(15F4.3)
1000
 2931000
 278 3771000
 109 379-0681000
 656 300 211 1291000
 825 331 311 160 6771000
 519 569 329 321 595 5661000
 191 659 336 378 137 263 4951000
 146 400 677-042 185 211 397 4091000
 168 474 554 026 140 233 353 626 6571000
 123 406-026 857 186 185 388 408 071 0491000
 079 362-039 822 082 124 320 509-026 080 8061000
 014 487 165 394 027 096 345 675 115 344 316 4571000
 022 451 252 257 058 083 305 557 269 554 175 250 7791000
-056 272-006 550-033-001 216 439-059 073 475 662 732 4641000
ME
*
75.352 57.621 57.931 44.421 75.338 30.731 74.150 28.886 63.167 28.867
43.767 24.243 69.679 68.971 63.400
SD
*
14.524 11.014 9.479 16.347 8.853 5.019 8.656 4.873 10.701 5.319 11.073
7.087 9.828 11.655 14.643
SELECTION
1 5 6 2 8 3 9 10 4 11 12/
MO LX=FU PH=SY TD=SY,FI
FR LX(2,1) LX(3,1) LX(5,2) LX(7,3) LX(8,3) LX(10,4) LX(11,4)
FR LX(6,1)
FR TD(1,1) TD(2,2) TD(3,3) TD(4,4) TD(5,5) TD(6,6) TD(7,7) TD(8,8)
FR TD(9,9) TD(10,10) TD(11,11)
FR TD(8,5) TD(10,7) TD(11,5)
ST 30.0 LX(1,1) LX(4,2) LX(6,3) LX(9,4)
ST 1.0 PH(1,1) PH(2,2) PH(3,3) PH(4,4)
ST .05 PH(2,1) PH(3,1) PH(4,1) PH(3,2)
ST -.01 PH(4,3)
ST 40.0 TD(1,1) TD(2,2) TD(4,4) TD(6,6) TD(7,7) TD(9,9)
ST 10.0 TD(5,5) TD(8,8) TD(11,11) TD(10,7)
ST 18.0 TD(10,10)
ST 5.0 TD(8,5) TD(11,5) TD(3,3)
OU NS
```

3.2. The MO Card

Before we can proceed further in testing for invariance, we must learn additional LISREL specification language, which is needed for the MO card; this information is now summarized.

1. The specification of any matrix element in a current group, contains only two indices within parentheses (e.g., LX(2,1). To refer to a matrix element in some other group, however, there must be three indices within parentheses, where the first number refers to the number of the group. For example, the specification of LX(2,2,1) refers to the element LX(2,1) in the second group; LX(3,2,1), to the element LX(2,1) for the third group, and so on.

2. In defining equality constraints between groups, the parameter to be constrained is specified as free in the first group and as equated (EQ) to the first group for each of the other groups. For example, in testing for the invariance of LX(2,1) across three groups, the specification would be:

In group 1: FR LX(2,1).
In group 2: EQ LX(1,2,1) LX(2,1).
In group 3: EQ LX(1,2,1) LX(2,1).

According to this specification, LX(2,1) will be freely estimated for group 1 only; for groups 2 and 3, the value of LX(2,1) will be constrained equal to the value obtained for group 1.

3. To constrain an entire matrix (i.e., all elements in the matrix) invariant across groups, the matrix of interest can itself be specified as invariant (IN) on the MO card. That is to say, to constrain LX in group 2 equal to LX for group 1, we would specify LX = IN. It is worth noting, however, that this specification at the LX matrix level is identical to a specification wherein each element in the LX matrix for group 2 is specified as equivalent (EQ) to each element in the LX matrix for group 1 (as explained earlier). For example, in the case of our present data, the specification LX = IN is equivalent to the following specification with respect to individual elements of LX:

$$\text{EQ LX}(1,1,1) \ \text{LX}(1,1)$$
$$\text{EQ LX}(1,2,1) \ \text{LX}(2,1)$$
$$\downarrow$$
$$\text{EQ LX}(1,11,4) \ \text{LX}(11,4)$$

4. Additionally, if start values were included for the initial baseline model input, they may need to be increased in order to make them compatible with covariance, rather than correlation values. In the case of the present data, for example, the start values for the fixed λs were changed from 1.00 to 30.00, and the free λs from 0.90 to 15.0.

5. Finally, the following matrix specifications can be described on the MO card:

SP—indicates that the matrix has the same pattern of fixed and free elements as the corresponding matrix in the previous group.

SS—indicates that the matrix will be given the same starting values as the corresponding matrix in the previous group.

PS—indicates that the same pattern and starting values will be used as in the corresponding matrix of the previous group.

3.3. The OU Card

Specification on this card remains consistent with other models described in Section II. In other words, whether or not the analyses are based on single or multiple groups has no bearing on the requested output.

4. Testing Hypotheses Related to Factorial Invariance

Now we are ready to test the series of hypotheses related to the invariance of SC measurements and structure across gender. Let's proceed by examining, one at a time, the model specification input and results associated with each of these hypotheses. For purposes of demonstration, the omnibus test of invariant variance-covariance matrices will be conducted. However, you are urged to refer to the earlier discussion regarding the rationale and problems associated with this test.

4.1. Hypothesis 1 (H_o: $\Sigma_M = \Sigma_F$)

The first hypothesis to be tested is the global one that relates to the equivalence of the variance-covariance matrices across sex. As noted by Jöreskog (1971a), one hopes to be able to reject this hypothesis in order to argue that the covariance matrices for males and females are significantly different. The input for this model, related only to the MO card (all other input remains the same) is as follows:

Males: MO NX = 11 NK = 11 LX = ID TD = ZE
Females: MO PH = IN[4]

For males, the model specified here considers each measure to represent one factor (i.e., an 11-factor model), with the factor loading matrix specified as an identity matrix (i.e., each measure has exactly one fixed loading for exactly one factor), and the error variance-covariance matrix, a zero matrix (i.e., a null matrix). The model specification for females is exactly the same, with the added restriction that the factor variance-covariance matrix is invariant.

The hypothesis of invariant covariance matrices was rejected (χ^2 (66) = 138.80, $p<0.001$). These results imply that for adolescent males and females, SC structure differs with respect to: the number of factors underlying SC, the pattern of factor loadings (i.e., SC measurements of the SC factors under study), and/or the variance of the SC factors and/or their covariances (i.e., relations among the SC factors under study). We proceed now to test hypotheses related to each of these possibilities by testing a series of increasingly restrictive models.

4.2. Hypothesis 2 (H_o:$\Lambda_{M,K = 4} = \Lambda_{F,K = 4}$)

The second hypothesis to be tested is that the number of factors underlying the SC structure, as postulated in Figure 6.1, is invariant across sex (i.e., a four-factor structure). The decision to reject or not to reject this hypothesis is based on the overall goodness-of-fit between the simultaneous model and the data; a satisfactory fit arguing for a factor structure that includes the same number of factors in each group. Unlike single-group analyses, multigroup analyses yield only one overall χ^2 value, albeit separate GFI and RMR measures for each group. Since the χ^2 values are

---

[4]Where model specifications for the second and all subsequent groups are the same as the first group, these specifications need not be included in the MO card (i.e., for females, NX = 11 NK = 11 LX = ID TD = ZE).

summative across groups, the value obtained in the multigroup specification should equal the sum of χ^2 values obtained for the single-group analyses. The model specification input for testing this hypothesis, then, is
identical to that presented in Table 6.1. As such, parameter specification
is consistent with the baseline model for each group; the only difference
is the simultaneous, rather than the separate, estimation of parameters.

The results of this analysis yielded a χ^2 (65) = 138.26; this value, of
course, is equal to the sum of the separate goodness-of-fit indices for
males (χ^2 (31) = 60.96) and females (χ^2 (34) = 77.30). As such, the four-
factor solution is considered to represent a substantively reasonable fit to
the data (χ^2/df = 2.13; BBI = 0.98). As a case in point regarding the
sensitivity of χ^2 to trivial differences between comparative models, it is
worthwhile to note that on the basis of statistical criteria, this model
would be rejected ($p<0.001$). Interestingly, McGaw and Joreskog (1971),
presented with similar findings, argued for an invariant number of factors
(i.e., nonrejection of Hypothesis 2) based on a Tucker-Lewis index equal
to 0.94.[5] Given the substantially poorer model fit in their data, it seems
justifiable, and indeed reasonable here, to conclude that a four-factor
structure underlies the data in the present application.

Nonetheless, while these results suggest that for both males and females, the data are fairly well described by general SC, academic SC,
English SC, and mathematics SC, they do not necessarily imply that the
actual factor loadings are the same across sex. Thus, the hypothesis of
an invariant pattern of factor loadings remains to be tested; we turn now
to this issue.

4.3. Hypothesis 3 (H_o: $\Lambda_M = \Lambda_F$)

The hypothesis to be tested here argues that all measurement scaling units
(i.e., the factor loadings) for each SC factor, as specified in Figure 6.2 are
equivalent across sex (except for the additional cross-loading for males
[λ_{93}]). In other words, the pattern of factor loadings is invariant. A more
explicit expansion of this hypothesis is shown in Figure 6.3.

The testing of Hypothesis 3, as well as subsequent hypotheses related
to invariance, is identical to those used in the model-fitting procedures
demonstrated in Chapters 3 and 4. That is, a model is estimated in which
the χ^2 parameters are constrained to be equal across gender; the resulting
goodness-of-fit is then compared with that of a less restrictive model in
which the same parameters are free to take on any value. The tenability
of the hypothesis rests on the statistical significance of the $\Delta\chi^2$, between
the two models.

In specifying this model, the input for the female group only is of inter-

---

[5]As with the BBI, the Tucker-Lewis Index ranges from 0.0 to 1.00, values closest
to 1.00 representing the best fit to the data.

$$
\Lambda_M =
\begin{bmatrix}
1.0 & 0 & 0 & 0 \\
\lambda_{21} & 0 & 0 & 0 \\
\lambda_{31} & 0 & 0 & 0 \\
0 & 1.0 & 0 & 0 \\
0 & \lambda_{52} & 0 & 0 \\
\lambda_{61} & 0 & 1.0 & 0 \\
0 & 0 & \lambda_{73} & 0 \\
0 & 0 & \lambda_{83} & 0 \\
0 & 0 & 0 & 1.0 \\
0 & 0 & 0 & \lambda_{10,4} \\
0 & 0 & 0 & \lambda_{11,4}
\end{bmatrix}
=
\begin{bmatrix}
1.0 & 0 & 0 & 0 \\
\lambda_{21} & 0 & 0 & 0 \\
\lambda_{31} & 0 & 0 & 0 \\
0 & 1.0 & 0 & 0 \\
0 & \lambda_{52} & 0 & 0 \\
\lambda_{61} & 0 & 1.0 & 0 \\
0 & 0 & \lambda_{73} & 0 \\
0 & 0 & \lambda_{83} & 0 \\
0 & 0 & 0 & 1.0 \\
0 & 0 & 0 & \lambda_{10,4} \\
0 & 0 & 0 & \lambda_{11,4}
\end{bmatrix}
= \Lambda_F
$$

FIGURE 6.3. Hypothesized Invariance of the Pattern of Factor Loadings Across Gender.

est. This is because in our stacked data setup, the female data are presented in the second group; by convention, the second group is constrained equal to the first group (i.e., model specifications for males remain consistent with those presented in Table 6.1). Since there is one factor loading that is already known to be noninvariant across the two groups (λ_{93}), we cannot constrain the entire Λ matrix invariant by using the specification LX = IN. We thus specify a model that constrains all λs except λ_{93}. The LISREL input, for females only, is presented in Table 6.2.

To recapitulate, the hypothesis of an invariant pattern of factor loadings was tested by constraining all λ parameters except λ_{93} to be equal, and then comparing this model (Model 2) with Model 1 in which only the number of factors was held invariant. The fit of our constrained model yielded a χ^2 (73) = 145.37. Since the difference in χ^2 was not significant ($\Delta\chi^2$ (8) = 7.11), the hypothesis of an invariant pattern of factor loadings was considered tenable. If, however, this hypothesis had been rejected, the next step would have been to proceed by testing, independently, the invariance of each factor loading (λ) in the factor-loading matrix (Λ). (This

TABLE 6.2. LISREL Input for Females: Testing for the Invariance of Factor Loadings (i.e., self-concept measurements)

```
TESTING EJJALITY - FEMALES
DA NO=420
LA
*
'SDQGSC' 'SDQASC' 'SDQESC' 'SDQMSC' 'APIGSC' 'SESGSC' 'APIASC' 'SCAASC'
'APIESC' 'SCAESC' 'APIMSC' 'SCAMSC' 'GPA' 'ENG' 'MATH'
KM SY
(15F4.3)
1000
 2931000
 278 3771000
 109 379-0681000
 656 300 211 1291000
 825 331 311 160 6771000
 519 569 329 321 595 5061000
 191 659 336 378 137 263 4951000
 146 400 677-042 185 211 397 4091000
 168 474 554 026 140 233 353 626 6571000
 123 406-026 857 186 185 386 408 071 0491000
 079 362-039 822 082 124 320 509-026 080 8061000
 014 487 165 394 027 096 345 675 115 344 316 4671000
 022 451 252 257 058 083 305 557 269 554 175 250 7791000
-056 272-006 550-033-001 216 439-059 073 475 662 732 4641000
ME
*
 75.302 57.021 57.931 44.421 75.338 30.731 74.150 28.866 63.107 28.867
 43.767 24.243 69.679 68.971 63.400
SD
*
 14.524 11.014 9.479 16.347 8.653 5.013 8.656 4.873 10.701 5.319 11.073
 7.087 9.828 11.655 14.643
SELECTION
1 5 6 2 8 3 9 10 4 11 12/
MO LX=FU PH=SY TD=SY,FI
FR LX(2,1) LX(3,1) LX(5,2) LX(7,3) LX(8,3) LX(10,4) LX(11,4)
FR LX(6,1)
FR TD(1,1) TD(2,2) TD(3,3) TD(4,4) TD(5,5) TD(6,6) TD(7,7) TD(8,8)
FR TD(9,9) TD(10,10) TD(11,11)
FR TD(8,5) TD(10,7) TD(11,5)
EQ LX(1,2,1) LX(2,1)
EQ LX(1,3,1) LX(3,1)
EQ LX(1,5,2) LX(5,2)
EQ LX(1,7,3) LX(7,3)
EQ LX(1,8,3) LX(8,3)
EQ LX(1,10,4) LX(10,4)
EQ LX(1,11,4) LX(11,4)
EQ LX(1,6,1) LX(6,1)
ST 30.0 LX(1,1) LX(4,2) LX(6,3) LX(9,4)
ST 1.0 PH(1,1) PH(2,2) PH(3,3) PH(4,4)
ST .05 PH(2,1) PH(3,1) PH(4,1) PH(3,2)
ST -.01 PH(4,3)
ST 40.0 TD(1,1) TD(2,2) TD(4,4) TD(6,6) TD(7,7) TD(9,9)
ST 10.0 TD(5,5) TD(8,8) TD(11,11) TD(10,7)
ST 18.0 TD(10,10)
ST 5.0 TD(8,5) TD(11,5) TD(3,3)
OU NS
```

technique is demonstrated next with respect to the SC factor variance-covariance matrix.)

4.4. Hypothesis 4 (H_o: $\Phi_M = \Phi_F$)

The hypothesis to be tested here bears on the structure of the SC construct by focusing on the invariance of the SC factor variance-covariance matrix across sex. Specifically, it argues for the group equivalence of the variance associated with each SC facet and relations (i.e., covariances) among the SC facets. As noted earlier, hypotheses related to invariance

TABLE 6.3. LISREL Input for Females: Testing for the Invariance of the Factor Variance/Covariance Matrices (i.e., self-concept structure)

```
TESTING EQUALITY - FEMALES
DA NO=420
LA
*
'SDGGSC' 'SDQASC' 'SDSESC' 'SDQMSC' 'APIGSC' 'SESESC' 'APIASC' 'SCAASC'
'APIESC' 'SCAESC' 'APIMSC' 'SCAMSC' 'GPA' 'ENG' 'MATH'
KM SY
(15F4.3)
1000
2931000
278 3771000
109 379-0631000
656 300 211 1291000
825 331 311 160 6771000
519 569 329 321 595 5661000
191 659 336 378 137 263 4951000
146 400 677-042 185 211 397 4091000
168 474 554 026 140 233 353 626 6571000
123 406-026 857 186 185 338 408 071 0491000
079 362-039 822 082 124 320 509-026 080 8061000
014 487 165 394 027 096 345 675 115 344 316 4671000
022 451 252 257 058 083 305 557 269 554 175 250 7791000
-056 272-006 550-033-001 216 439-059 073 475 652 732 4641000
ME
*
75.352 57.021 57.931 44.421 75.338 30.731 74.150 28.886 63.107 28.807
43.767 24.243 69.679 68.971 63.400
SD
*
14.524 11.014 9.479 16.347 8.853 5.019 8.656 4.873 10.701 3.319 11.073
7.087 9.828 11.655 14.643
SELECTION
1 5 6 2 8 3 9 10 4 11 12/
MO LX=FU PH=IN TD=SY,FI
FR LX(2,1) LX(3,1) LX(5,2) LX(7,3) LX(8,3) LX(10,4) LX(11,4)
FR LX(6,1)
FR TD(1,1) TD(2,2) TD(3,3) TD(4,4) TD(5,5) TD(5,6) TD(7,7) TD(8,8)
FR TD(9,9) TD(10,10) TD(11,11)
FR TD(8,5) TD(10,7) TD(11,3)
EQ LX(1,2,1) LX(2,1)
EQ LX(1,3,1) LX(3,1)
EQ LX(1,5,2) LX(5,2)
EQ LX(1,7,3) LX(7,3)
EQ LX(1,8,3) LX(8,3)
EQ LX(1,10,4) LX(10,4)
EQ LX(1,11,4) LX(11,4)
EQ LX(1,6,1) LX(6,1)
ST 30.0 LX(1,1) LX(4,2) LX(6,3) LX(9,4)
ST 40.0 TD(1,1) TD(2,2) TD(4,4) TD(6,6) TD(7,7) TD(9,9)
ST 10.0 TD(5,5) TD(8,8) TD(11,11) TD(10,7)
ST 18.0 TD(10,10)
ST 5.0 TD(8,5) TD(11,5) TD(3,3)
OU NS
```

involve increasingly restrictive models; as such, the model to be tested here (Model 3) is more restrictive than Model 2. In addition to the specification of constraints between λs, Model 3 also includes the restriction that the entire factor variance-covariance matrix (Φ) be constrained invariant across sex (PH = IN). The LISREL input for Model 3, again for females only, is presented in Table 6.3.

The fit of Model 3 with the data yielded a χ^2 (83) = 195.30. Since the difference in χ^2 between this model and Model 2 was statistically significant ($\Delta\chi^2$ (10) = 49.93, $p<0.001$), the hypothesis of equivalent SC structure was rejected. The strategy at this point was to isolate those components of SC structure that were noninvariant across sex. To determine

TABLE 6.4. LISREL Input for Females: Testing for the Invariance of Individual
Factor Variance/Covariance Parameters

```
TESTING EQUALITY - FEMALES
DA NO=420
LA
*
'SDQGSC' 'SDQASC' 'SDQESC' 'SDQMSC' 'APISSC' 'SESGSC' 'APIASC' 'SCAASC'
'APIESC' 'SCAESC' 'APIMSC' 'SCAMSC' 'GPA' 'ENG' 'MATH'
KM SY
(15F4.3)
1000
 2931000
  278 3771000
  109 379-068 1000
  656 300 211 1291000
  825 331 311 160 6771000
  519 569 329 321 595 5661000
  191 659 336 378 137 263 4951000
  146 400 677-042 185 211 397 4091000
  168 474 554 026 140 233 393 626 6571000
  123 406-026 857 186 185 388 408 071 0491000
  079 362-039 822 082 124 320 509-026 080 8061000
  014 487 165 394 027 096 345 675 115 344 316 4671000
  022 451 252 257 058 083 305 557 269 554 175 250 7791000
 -056 272-006 550-033-001 216 439-059 073 475 662 732 4641000
ME
*
 75.352 57.021 57.931 44.421 75.338 30.731 74.150 28.886 63.167 28.867
 43.767 24.243 69.679 68.971 63.400
SD
*
 14.524 11.014 9.479 16.347 8.853 5.019 8.656 4.873 10.701 5.319 11.073
 7.087 9.828 11.655 14.643
SELECTION
1 5 6 2 8 3 9 10 4 11 12/
MO LX=FU PH=SY TD=SY,FI
FR LX(2,1) LX(3,1) LX(5,2) LX(7,3) LX(8,3) LX(10,4) LX(11,4)
FR LX(6,1)
FR TD(1,1) TD(2,2) TD(3,3) TD(4,4) TD(5,5) TD(6,6) TD(7,7) TD(8,8)
FR TD(9,9) TD(10,10) TD(11,11)
FR TD(8,5) TD(10,7) TD(11,5)
EQ LX(1,2,1) LX(2,1)
EQ LX(1,3,1) LX(3,1)
EQ LX(1,5,2) LX(5,2)
EQ LX(1,7,3) LX(7,3)
EQ LX(1,8,3) LX(8,3)
EQ LX(1,10,4) LX(10,4)
EQ LX(1,11,4) LX(11,4)
EQ LX(1,6,1) LX(6,1)
EQ PH(1,1,1) PH(1,1)
ST 30.0 LX(1,1) LX(4,2) LX(6,3) LX(9,4)
ST 1.0 PH(1,1) PH(2,2) PH(3,3) PH(4,4)
ST .05 PH(2,1) PH(3,1) PH(4,1) PH(3,2)
ST -.01 PH(4,3)
ST 40.0 TD(1,1) TD(2,2) TD(4,4) TD(6,6) TD(7,7) TD(9,9)
ST 10.0 TD(5,5) TD(8,8) TD(11,11) TD(10,7)
ST 18.0 TD(10,10)
ST 5.0 TD(8,5) TD(11,5) TD(3,3)
OU NS
```

this information, it was necessary to test, independently, the invariance
of each parameter in the phi matrix. As such, Model 4 was specified with
the variance of general SC (ϕ_{11}) constrained equal [EQ PH(1,1,1)
PH(1,1)], in addition to the parameters known to be invariant across the
groups. The LISREL input, for females only, is presented in Table 6.4.

Since the difference in fit between this model (Model 4) and Model 2,
in which only the factor loadings were held invariant, was not significant
($\Delta\chi^2$ (1) = 1.15), the hypothesis of equivalent variance in general SC was
considered tenable. In like manner, a series of subsequent models were

TABLE 6.5. Simultaneous Tests for the Invariance of Self-Concept Measurements and Structure

Competing models	χ^2	df	$\Delta\chi$	df	χ^2/df	BBI
0 Null model	6,465.41	110	—	—	—	.97
1 Number of factors invariant	138.26	65	—	—	2.13	.98
2 Number of factors and pattern of loadings invariant[a]	145.37	73	7.11	8	1.99	.98
3 Model 2 with all latent variances and covariances invariant	195.30	83	49.93[**]	10	2.35	.97
4 Model 2 with latent construct parameters made independently invariant						
Variances						
(a) General SC	146.52	74	1.15	1	1.98	.98
(b) Academic SC	161.72	74	16.35[**]	1	2.19	.98
(c) English SC	147.17	74	1.80	1	1.99	.98
(d) Mathematics SC	146.16	74	.79	1	1.98	.98
Covariances						
(e) Academic/general SC	148.55	74	3.18	1	2.01	.98
(f) English/general SC	145.37	74	0.00	1	1.96	.98
(g) Mathematics/general SC	149.54	74	4.17[*]	1	2.02	.98
(h) English/academic SC	146.89	74	1.52	1	1.99	.98
(i) Mathematics/academic SC	167.39	74	22.02[**]	1	2.26	.97
(j) Mathematics/English SC	157.56	74	12.19[**]	1	2.13	.98

[*]$p<0.05$ [**]$p<0.001$
[a]All lambda parameters invariant except λ_{93}.
From Byrne and Shavelson (1987), "Adolescent Self-concept: Testing the Assumption of Equivalent Structure Across Gender" in *American Educational Research Journal, 24(3),* 365–385. Copyright 1987 by American Educational Research Association. Reprinted with permission.

specified in which each of the remaining parameters in the Φ matrix was constrained equal across sex. Overall, tests of hypotheses related to the equivalence of SC structure revealed significant gender differences in the variance of academic SC, and with respect to relations between: mathematics SC and general SC, mathematics SC and academic SC, and mathematics SC and English SC. A summary of the testing of all hypotheses related to the invariance of SC measurements and structure across gender is presented in Table 6.5.

Given the differences between males and females illustrated in Table 6.5, invariance of the error variance-covariance matrices was not formally tested. Although Jöreskog (1971a) suggested that the hypothesis of an invariant variance-covariance matrix (Φ) be tested conditional on the findings of an invariant number of factors, factor-loading pattern, and error variances, this restriction is excessively stringent and not always necessary (Muthén, personal communication, Jan. 1986; Alwin & Jackson, 1980).

5. Summary

This chapter demonstrated how to test hypotheses related to factorial invariance across groups. Specifically, procedures were demonstrated that tested for the gender invariance of a multidimensional adolescent SC structure as measured by multiple measuring instruments. The first step was to fit a hypothesized four-factor model of SC separately for males and females. We next combined the two baseline models and proceeded to test hypotheses related to the invariance of SC measures (i.e., factor loadings) and SC structure (i.e., factor variances and covariances). Finally, procedures used in testing for and with partial measurement invariance were illustrated.

7
Testing for Item Invariance of a Measuring Instrument

In this second multigroup application, our attention focuses on invariance as it relates to a single measuring instrument, the Self Description Questionnaire (SDQIII; Marsh & O'Neill, 1984). In this chapter we explore the factorial equivalency of the SDQIII across two academically tracked (low and high) groups of high school students. (For a more extensive discussion of academic tracking as it relates to this data base, see Byrne, 1988a; for details of the study related to this application, see Byrne, 1988c.)

Typically, in testing for the factorial invariance of a single measuring instrument, the researcher is primarily interested in three psychometric issues: that the items comprising each subscale are factorially valid and equivalent across groups, that the factor covariances (i.e., relations among the underlying construct dimensions) are equivalent across groups, and that the subscale items are equally reliable across groups.

This chapter addresses each of these issues. Specifically, it illustrates the testing of hypotheses bearing on the equivalency of the SDQIII across academic track; consistent with the procedures outlined in Chapter 4, all analyses are based on item-pairs. The reader is encouraged to refer to Chapter 4 for a review of other details regarding a description of the SDQIII, the hypothesized model, and procedures for establishing a baseline model for each group.

1. Tests for Invariance Related to the SDQIII

1.1. The Hypothesized Model

As in Chapter 4, all analyses were based on item-pairs, rather than on single items. Likewise, the CFA model under study hypothesized a priori that: responses to the SDQIII could be explained by four factors (general SC, academic SC, English SC, and mathematics SC), each item-pair would have a nonzero loading on the SC factor it was designed to measure (i.e., the target factor) and zero loadings on all other factors (i.e., nontarget factors), the four factors would be correlated, and the uniquenesses for the item-pair variables would be uncorrelated. This model was presented schematically in Figure 4.1, and the pattern of parameters to be estimated was detailed in Table 4.1.

1.2. The Baseline Models

Although for both tracks the hypothesized four-factor model represented a psychometrically reasonable fit to the data (low track, $\chi^2/df = 2.32$, BBI = 0.90; high track, $\chi^2/df = 4.40$, BBI = 0.91), the fit, based solely on statistical criteria, was less than adequate (low track, $\chi^2_{(183)} = 425.18$, $p<0.001$; high track $\chi^2_{(183)} = 805.45$, $p<0.001$).

To investigate the misfit in the model, a sensitivity analysis was conducted as outlined in Chapter 4. As such, model fitting for each track was continued beyond the initially fitted models. Several additional modifications that included both correlated uniquenesses and secondary loadings (item-pair loadings on nontarget factors), resulted in a statistically better fitting model for both the low track $\chi^2_{(164)} = 192.48$, $p = 0.06$; BBI = 0.94) and the high track ($\chi^2_{(149)} = 171.52, p = 0.10$; BBI = 0.98). Given the probability of method effects (see Byrne, 1988b; Gerbing & Anderson, 1984) and the moderate correlations among the four SC factors under study, these parameters were not unexpected.

Based on the following considerations, however, these final models were rejected in favor of the more parsimonious initial models: the uniqueness covariance estimates, while statistically significant, were relatively minor (low track $\overline{X} = 0.05$; high track, $\overline{X} = 0.04$); the estimated secondary factor loadings, while statistically significant, were also relatively minor (low track, $\overline{X} = 0.04$; high track, $\overline{X} = 0.03$); the estimated factor loading and factor variance-covariance estimates in the final model correlated 0.93 and 0.99, respectively, for the low track, and 0.94 and 0.97, respectively, for the high track, with those in the initially hypothesized model (see Byrne et al., 1989; Tanaka & Huba, 1984), these results substantiating the stability of the initial models; although each of the model respecifications yielded a statistically significant improvement in model fit, these increments based on the BBI could be considered of little practical importance (see also, Marsh & Hocevar, 1985); the sensitivity of the χ^2 likelihood ratio test with large samples is now widely known (see Bentler & Bonett, 1980; Marsh & Hocevar, 1985); and given the exploratory nature of these supplementary analyses, and thus the risk of capitalization on chance factors (see MacCallum, 1986), the final model estimates were considered dubious. For these reasons, then, the initial model for each track was used as the baseline model in tests of invariance; these models are presented schematically in Figure 7.1.

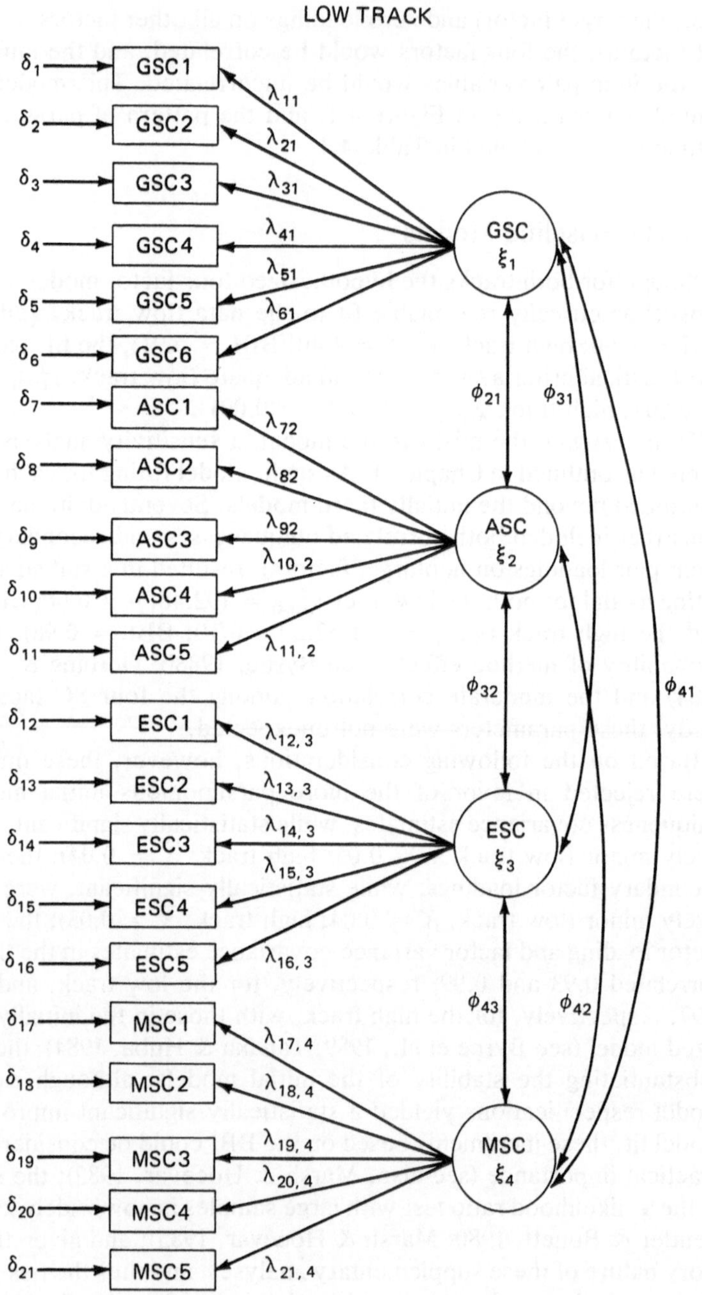
LOW TRACK
δ_1 GSC1 λ_{11}
δ_2 GSC2 λ_{21}
δ_3 GSC3 λ_{31}
δ_4 GSC4 λ_{41}
δ_5 GSC5 λ_{51}
δ_6 GSC6 λ_{61}
δ_7 ASC1 λ_{72}
δ_8 ASC2 λ_{82}
δ_9 ASC3 λ_{92}
δ_{10} ASC4 $\lambda_{10,2}$
δ_{11} ASC5 $\lambda_{11,2}$
δ_{12} ESC1 $\lambda_{12,3}$
δ_{13} ESC2 $\lambda_{13,3}$
δ_{14} ESC3 $\lambda_{14,3}$
δ_{15} ESC4 $\lambda_{15,3}$
δ_{16} ESC5 $\lambda_{16,3}$
δ_{17} MSC1 $\lambda_{17,4}$
δ_{18} MSC2 $\lambda_{18,4}$
δ_{19} MSC3 $\lambda_{19,4}$
δ_{20} MSC4 $\lambda_{20,4}$
δ_{21} MSC5 $\lambda_{21,4}$
GSC ξ_1
ASC ξ_2
ESC ξ_3
MSC ξ_4
ϕ_{21} ϕ_{31} ϕ_{41}
ϕ_{32} ϕ_{42} ϕ_{43}

FIGURE 7.1.

HIGH TRACK

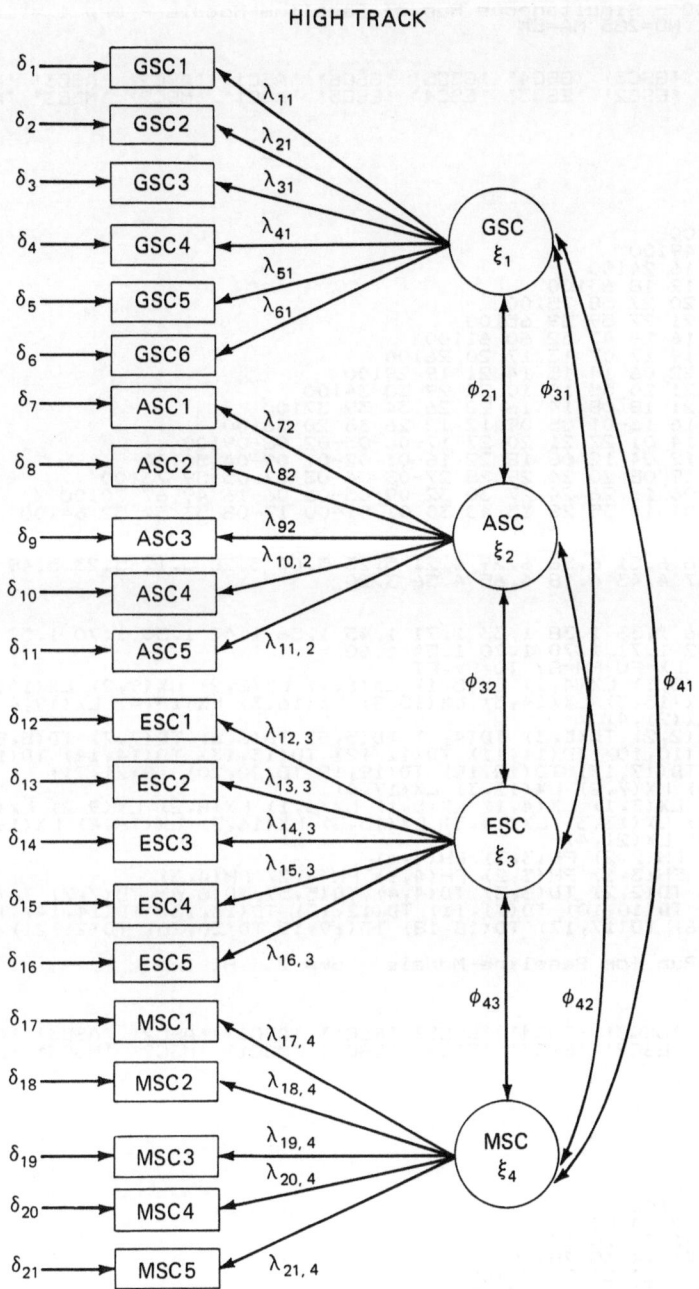

FIGURE 7.1. Continued.

```
CFA of the SDQ - Simultaneous Run of Baseline Models - Grp 1 = Lo Track
DA NG=2 NI=21 NO=285 MA=CM
LA
*
'GSC1' 'GSC2' 'GSC3' 'GSC4' 'GSC5' 'GSC6' 'ASC1' 'ASC2' 'ASC3' 'ASC4'
'ASC5' 'ESC1' 'ESC2' 'ESC3' 'ESC4' 'ESC5' 'MSC1' 'MSC2' 'MSC3' 'MSC4'
'MSC5'
KM SY
(21F3.2)
1000
 69100
 70 68100
 53 50 44100
 65 61 66 53100
 53 51 53 51 49100
 18 18 13 20 16 24100
 19 14 13 18 12 18 63100
 20 17 17 22 20 27 58 75100
 25 26 23 27 21 27 59 59 65100
 22 23 15 22 16 29 43 52 60 61100
 17 15 12 11 19 17 07 13 17 20 26100
 16 16 14 25 22 06 14 15 14 21 19 29100
 13 22 17 16 21 20 05 10 10 22 29 38 34100
 19 12 21 15 21 18 08 14 16 28 26 34 39 37100
 15 13 09 29 16 16-01 05 09 12 12 26 33 20 26100
 13 17 17 13 14 01 27 21 20 27 17-06-05-02 08-09100
 15 15 16 14 12 04 12 08 18 22 16-01-02-01 09-04 50100
 25 17 20 15 19 08 20 24 29 28 27-03 04-03 07-05 59 73100
 13 11 11 08 06 11 26 24 27 30 32-09-05-06 06-16 49 67 70100
 16 14 16 05 01 14 35 28 35 43 30 01 01-00 17-08 53 57 52 64100
ME
*
6.14 6.52 5.96 6.51 6.56 6.29 4.21 5.23 5.00 5.28 5.12 5.23 5.49
5.71 5.31 6.67 4.45 4.18 4.65 4.56 3.20
SD
*
1.42 1.25 1.36 1.33 1.38 1.33 1.71 1.45 1.56 1.47 1.50 1.70 1.52
1.41 1.46 1.82 1.71 1.70 1.70 1.54 1.60
MO NX=21 NK=4 LX=FU PH=SY TD=SY,FI
FR LX(2,1) LX(3,1) LX(4,1) LX(5,1) LX(6,1) LX(8,2) LX(9,2) LX(10,2)
FR LX(11,2) LX(13,3) LX(14,3) LX(15,3) LX(16,3) LX(18,4) LX(19,4)
FR LX(20,4) LX(21,4)
FR TD(1,1) TD(2,2) TD(3,3) TD(4,4) TD(5,5) TD(6,6) TD(7,7) TD(8,8)
FR TD(9,9) TD(10,10) TD(11,11) TD(12,12) TD(13,13) TD(14,14) TD(15,15)
FR TD(16,16) TD(17,17) TD(18,18) TD(19,19) TD(20,20) TD(21,21)
ST 1.0 LX(1,1) LX(7,2) LX(12,3) LX(17,4)
ST .7 LX(2,1) LX(3,1) LX(4,1) LX(5,1) LX(6,1) LX(8,2) LX(9,2) LX(10,2)
ST .7 LX(11,2) LX(13,3) LX(14,3) LX(15,3) LX(16,3) LX(18,4) LX(19,4)
ST .7 LX(20,4) LX(21,4)
ST .5 PH(1,1) PH(2,2) PH(3,3) PH(4,4)
ST .3 PH(2,1) PH(3,1) PH(3,2) PH(4,1) PH(4,2) PH(4,3)
ST .2 TD(1,1) TD(2,2) TD(3,3) TD(4,4) TD(5,5) TD(6,6) TD(7,7) TD(8,8)
ST .2 TD(9,9) TD(10,10) TD(11,11) TD(12,12) TD(13,13) TD(14,14) TD(15,15
ST .2 TD(16,16) TD(17,17) TD(18,18) TD(19,19 TD(20,20) TD(21,21)
OU NS
Simultaneous Run for Baseline Models - Grp 2 = Hi Track
DA NO=613
LA
*
'GSC1' 'GSC2' 'GSC3' 'GSC4' 'GSC5' 'GSC6' 'ASC1' 'ASC2' 'ASC3' 'ASC4'
'ASC5' 'ESC1' 'ESC2' 'ESC3' 'ESC4' 'ESC5' 'MSC1' 'MSC2' 'MSC3' 'MSC4'
'MSC5'
KM SY
(21F3.2)
1000
 71100
 79 69100
 63 57 62100
 74 71 77 69100
 62 59 61 68 64100
 25 24 19 28 19 24100
 21 22 18 30 26 23 70100
 23 22 23 31 21 24 55 78100
 25 22 24 33 23 24 50 70 76100
 25 23 21 31 21 29 42 61 72 69100
 15 04 14 09 10 08 14 22 24 26 23100
 13 11 13 14 14 07 18 29 28 33 23 41100
 25 18 31 21 25 22 12 24 30 35 28 42 44100
 33 24 32 21 27 23 19 27 30 42 32 45 47 42100
 04 01 03 10 08 05 08 19 19 21 21 19 35 24 24100
 14 11 09 17 12 14 34 39 34 36-06 02-02 08 11100
 11 12 12 13 11 14 18 22 25 26 35-15-06-08 02 02 71100
 11 11 11 10 11 15 21 32 35 35 42-10 01-00 06 08 75 81100
 12 11 11 15 08 15 25 34 39 40 47-05 05 05 11 08 69 78 82100
 08 07 07 15 09 13 28 33 32 34 41-06 04 02 10 05 67 72 72 74100
```

TABLE 7.1. Continued

```
ME
*
6.21  6.40  5.96  6.47  6.55  6.17  4.66  5.96  5.90  6.02  6.17  5.52
5.81  5.92  5.50  7.10  5.30  4.64  5.29  5.19  4.13
SD
*
1.38  1.30  1.39  1.36  1.48  1.39  1.58  1.33  1.38  1.30  1.40  1.71
1.62  1.31  1.43  1.52  1.88  2.05  1.86  1.78  1.82
MO NX=21 NK=4 LX=FU PH=SY TD=SY,FI
FR LX(2,1) LX(3,1) LX(4,1) LX(5,1) LX(6,1) LX(8,2) LX(9,2) LX(10,2)
FR LX(11,2) LX(13,3) LX(14,3) LX(15,3) LX(16,3) LX(18,4) LX(19,4)
FR LX(20,4) LX(21,4)
FR TD(1,1) TD(2,2) TD(3,3) TD(4,4) TD(5,5) TD(6,6) TD(7,7) TD(8,8)
FR TD(9,9) TD(10,10) TD(11,11) TD(12,12) TD(13,13) TD(14,14) TD(15,15)
FR TD(16,16) TD(17,17) TD(18,18) TD(19,19 TD(20,20) TD(21,21)
ST 1.0 LX(1,1) LX(7,2) LX(12,3) LX(17,4)
ST .7 LX(2,1) LX(3,1) LX(4,1) LX(5,1) LX(6,1) LX(8,2) LX(9,2) LX(10,2)
ST .7 LX(11,2) LX(13,3) LX(14,3) LX(15,3) LX(16,3) LX(18,4) LX(19,4)
ST .7 LX(20,4) LX(21,4)
ST .5 PH(1,1) PH(2,2) PH(3,3) PH(4,4)
ST .3 PH(2,1) PH(3,1) PH(3,2) PH(4,1) PH(4,2) PH(4,3)
ST .2 TD(1,1) TD(2,2) TD(3,3) TD(4,4) TD(5,5) TD(6,6) TD(7,7) TD(8,8)
ST .2 TD(9,9) TD(10,10) TD(11,11) TD(12,12) TD(13,13) TD(14,14) TD(15,15)
ST .2 TD(16,16) TD(17,17) TD(18,18) TD(19,19 TD(20,20) TD(21,21)
OU NS
```

2. Tests for Invariance Across Ability

2.1. Testing for Equivalent Item-Pair Measurements

Given Muthén's comments regarding the omnibus test of invariant variance-covariance matrices (see Chapter 6), we begin by first testing for an invariant number of factors across track. As such, the high- and low-track baseline models were combined into one file and the model specified as a two-group four-factor model. The fit of this four-factor solution, which we will call Model 1, yielded a reasonable fit to the data ($\chi^2_{(366)}$ = 1230.64; BBI = 0.90).[1] On the basis of these results, we conclude that for both tracks the data are adequately described by the four hypothesized factors of SC. The LISREL input for this simultaneous run (Model 1), is presented in Table 7.1. The presentation of specification data for Model 1 in two different formats is intended to assist you in making the correspondence between the model schema as presented in Figure 7.1 and the computer input as presented in Table 7.1.

A couple of points related to the Model 1 input are worthy of note. First, the theta-delta matrix could also have been specified as TD = DI. As such, there would be no need to specify all θ_δ parameters as free; start values, however, would be specified for each of these parameters. Second, you will note that, unlike the application presented in Chapter 6, the start values here remained unchanged from the single-group analyses based on the correlation matrix (KM) and the multiple-group analyses based on the covariance matrix (CM). The reason for leaving the start

---

[1] Recall that this goodness-of-fit index represents the sum of the χ^2 values for the separately estimated baseline models.

values intact here was because a convergent solution was achieved with the original values.[2]

As noted in Chapter 6, evidence that the data are well described by four SC factors for both the low and the high tracks (see Model 1) in no way implies that the actual factor loadings are the same across track; this hypothesis must be tested and is done so by placing equality constraints on all parameters. The LISREL input for this model (Model 2), as specified for the high track (group 2) only, is illustrated in Table 7.2.[3] Of particular importance in the MO statement is the specification that the Λ matrix be constained invariant across groups (LX = IN). It is worth noting that although the specification of λ parameters as free or fixed has been deleted from the input, as have the start values for these parameters, this need not have been done. Had these specifications been included, however, the imposition of a constrained Λ matrix would have caused LISREL to override these commands.

The hypothesis of an invariant pattern of factor loadings across track was found to be tenable ($\Delta\chi^2$ (17) = 23.91); the $\Delta\chi^2$ representing the difference in model fit between Model 1 (number of factors constrained equal) and Model 2 (number of factors and pattern of factor loadings constrained equal). From this information we can conclude that all items comprising the four subscales of the SDQ are measuring the same SC facet in the same scaling units for both the low and high tracks.

It is important to note that if, on the other hand, the hypothesis of an equivalent pattern of factor loadings had been rejected, we would want to investigate further, the source of this noninvariance. Therefore, we would proceed to test, independently, each item-pair factor-loading parameter in the matrix. (This technique will be demonstrated in Chapter 8.)

2.2. Testing for Equivalent Factor Covariances

Testing for the invariance of factor covariances bears on the group equivalence of SC relations as measured by the four SDQIII subscales. To test this hypothesis, equality constraints are imposed, independently, on each

----

[2]A nonconvergent solution is evidenced by the error message that LX is written on————KSI 2 (or something comparable). Like many other error messages in LISREL, this one bears no clue to the problem of inappropriate start values. Should you receive this message, however, the start values should be made larger, consistent with covariance, rather than correlation values (see Chapter 6).
[3]Since in testing for invariance using LISREL, the model of interest is specified such that constraints on the model (i.e., specification of equalities across groups) are specified on the MO card for the second and subsequent groups only, the model specification for group 1 therefore remains intact and is never altered. Thus, the LISREL input related to the MO and subsequent cards for group 2 only is of interest for all remaining tests for invariance.

TABLE 7.2. LISREL Specification Input for Model 2 (High Track Only): Testing for an Invariant Pattern of Factor Loadings

```
Testing for Invariance of Item Pairs - Grp 2 = Hi Track
DA NO=613
LA
*
'GSC1' 'GSC2' 'GSC3' 'GSC4' 'GSC5' 'GSC6' 'ASC1' 'ASC2' 'ASC3' 'ASC4'
'ASC5' 'ESC1' 'ESC2' 'ESC3' 'ESC4' 'ESC5' 'MSC1' 'MSC2' 'MSC3' 'MSC4'
'MSC5'
KM SY
(21F3.2)
100 0
 71100
 79 69100
 63 57 62100
 74 71 77 69100
 62 59 61 68 64100
 25 24 19 28 19 24100
 21 22 18 30 26 23 70100
 23 22 23 31 21 24 55 78100
 25 22 24 33 23 24 50 70 76100
 25 23 21 31 21 29 42 61 72 69100
 15 04 14 09 10 08 14 22 24 26 23100
 13 11 13 14 14 07 18 29 28 33 23 41100
 25 18 31 21 25 22 12 24 30 35 28 42 44100
 33 24 32 21 27 23 19 27 30 42 32 45 47 42100
 04 01 03 10 08 05 08 19 19 21 21 19 35 24 24100
 14 11 09 17 12 14 34 39 34 36 36-06 02-02 08 11100
 11 12 12 13 11 14 18 22 25 26 35-15-06-08 02 02 71100
 11 11 11 10 11 15 21 32 35 35 42-10 01-00 06 08 75 81100
 12 11 11 15 08 15 25 34 39 40 47-05 05 05 11 08 69 78 82100
 08 07 07 15 09 13 28 33 32 34 41-06 04 02 10 05 67 72 72 74100
ME
*
6.21 6.40 5.96 6.47 6.55 6.17 4.66 5.96 5.90 6.02 6.17 5.52
5.81 5.92 5.50 7.10 5.30 4.64 5.29 5.19 4.13
SD
*
1.38 1.30 1.39 1.36 1.48 1.39 1.58 1.33 1.38 1.30 1.40 1.71
1.62 1.31 1.43 1.52 1.88 2.05 1.86 1.78 1.82
MO LX=IN PH=SY TD=SY,FI
FR TD(1,1) TD(2,2) TD(3,3) TD(4,4) TD(5,5) TD(6,6) TD(7,7) TD(8,8)
FR TD(9,9) TD(10,10) TD(11,11) TD(12,12) TD(13,13) TD(14,14) TD(15,15)
FR TD(16,16) TD(17,17) TD(18,18) TD(19,19 TD(20,20) TD(21,21)
ST .5 PH(1,1) PH(2,2) PH(3,3) PH(4,4)
ST .3 PH(2,1) PH(3,1) PH(3,2) PH(4,1) PH(4,2) PH(4,3)
ST .2 TD(1,1) TD(2,2) TD(3,3) TD(4,4) TD(5,5) TD(6,6) TD(7,7) TD(8,8)
ST .2 TD(9,9) TD(10,10) TD(11,11) TD(12,12) TD(13,13) TD(14,14) TD(15,15)
ST .2 TD(16,16) TD(17,17) TD(18,18) TD(19,19 TD(20,20) TD(21,21)
OU NS
```

of these phi parameters (ϕ_{21}, ϕ_{31}, ϕ_{41}, ϕ_{32}, ϕ_{42}, ϕ_{43}). However, recall that tests for invariance are based on a series of successively specified models, such that each is more restrictive than the former; such models are referred to as "nested models." In keeping with this model-nesting mode then, we now increase the number of restrictions in Model 2 by adding equality constraints for each of these covariance parameters. This is accomplished by placing equality constraints on both the LX matrix and the particular parameters representing covariances; we'll call this Model 3. The LISREL specification input for this model (again for the high track only) is presented in Table 7.3. Note that since the entire Φ matrix was not constrained equal across groups (which would need to be indicated on the MO card), a separate equality constraint statement must be specified for each parameter of interest.

Since the fit differential between Models 2 and 3 is found to be nonsig-

TABLE 7.3. LISREL Specification Input for Model 3 (High Track Only): Testing for Invariant Factor Covariances

```
Testing for Invariance of Item Pairs - Grp 2 = Hi Track
DA NO=613
LA
*
'GSC1' 'GSC2' 'GSC3' 'GSC4' 'GSC5' 'GSC6' 'ASC1' 'ASC2' 'ASC3' 'ASC4'
'ASC5' 'ESC1' 'ESC2' 'ESC3' 'ESC4' 'ESC5' 'MSC1' 'MSC2' 'MSC3' 'MSC4'
'MSC5'
KM SY
(21F3.2)
1000
 71100
 79 69100
 63 57 62100
 74 71 77 69100
 62 59 61 68 64100
 25 24 19 28 19 24100
 21 22 18 30 26 23 70100
 23 22 23 31 21 24 55 78100
 25 22 24 33 23 24 50 70 76100
 25 23 21 31 21 29 42 61 72 69100
 15 04 14 09 10 08 14 22 24 26 23100
 13 11 13 14 14 07 18 29 28 33 23 41100
 25 18 31 21 25 22 12 24 30 35 28 42 44100
 33 24 32 21 27 23 19 27 30 42 32 45 47 42100
 04 01 03 10 08 05 08 19 19 21 21 19 35 24 24100
 14 11 09 17 12 14 34 39 34 36 36-06 02-02 08 11100
 11 12 12 13 11 14 18 22 25 26 35-15-06-08 02 02 71100
 11 11 11 10 11 15 21 32 35 35 42-10 01-00 06 08 75 81100
 12 11 11 15 08 15 25 34 39 40 47-05 05 05 11 08 69 78 82100
 08 07 07 15 09 13 28 33 32 34 41-06 04 02 10 05 67 72 72 74100
ME
*
6.21 6.40 5.96 6.47 6.55 6.17 4.66 5.96 5.90 6.02 6.17 5.52
5.81 5.92 5.50 7.10 5.30 4.64 5.29 5.19 4.13
SD
*
1.38 1.30 1.39 1.36 1.48 1.39 1.58 1.33 1.38 1.30 1.40 1.71
1.62 1.31 1.43 1.52 1.88 2.05 1.86 1.78 1.82
MO LX=IN PH=SY TD=SY,FI
FR TD(1,1) TD(2,2) TD(3,3) TD(4,4) TD(5,5) TD(6,6) TD(7,7) TD(8,8)
FR TD(9,9) TD(10,10) TD(11,11) TD(12,12) TD(13,13) TD(14,14) TD(15,15)
FR TD(16,16) TD(17,17) TD(18,18) TD(19,19 TD(20,20) TD(21,21)
ST .5 PH(1,1) PH(2,2) PH(3,3) PH(4,4)
ST .3 PH(2,1) PH(3,1) PH(3,2) PH(4,1) PH(4,2) PH(4,3)
ST .2 TD(1,1) TD(2,2) TD(3,3) TD(4,4) TD(5,5) TD(6,6) TD(7,7) TD(8,8)
ST .2 TD(9,9) TD(10,10) TD(11,11) TD(12,12) TD(13,13) TD(14,14) TD(15,15)
ST .2 TD(16,16) TD(17,17) TD(18,18) TD(19,19 TD(20,20) TD(21,21)
EQ PH(1,2,1) PH(2,1)
EQ PH(1,3,1) PH(3,1)
EQ PH(1,4,1) PH(4,1)
EQ PH(1,3,2) PH(3,2)
EQ PH(1,4,2) PH(4,2)
EQ PH(1,4,3) PH(4,3)
OU NS
```

nificant ($\Delta\chi^2$ (6) = 6.42), the hypothesis of invariant factor covariances is considered tenable. This finding provides evidence that the theoretical structure of SC, as measured by the SDQIII, is the same for both the low and high tracks. (Compare these results with those across gender using multiple measures of SC, as demonstrated in Chapter 6.)

Once again, if the hypothesis of equivalent factor covariances had been found untenable, we would be well advised to investigate further, the source of this noninvariance by testing, independently, each factor covariance parameter in the Φ matrix; model specification, of course, would include the invariant or partially invariant λ parameters.

2.3. Testing for Equivalent Item-Pair Reliabilities

Generally speaking, in multiple-indicator CFA models, testing for the invariance of reliability is neither necessary (Jöreskog, 1971a), nor of particular interest when the scales are used merely as CFA indicators and not as measures in their own right, ignoring reliability (Muthén, personal communication, Oct. 1987). Although Jöreskog (1971b) demonstrated the steps involved in testing for a completely invariant model (i.e., invariant Λ, Φ, and Θ), this procedure is considered an excessively stringent test of factorial invariance (Muthén, personal communication, Jan. 1987). In fact, Jöreskog (1971a) has shown that while it is necessary that multiple measures of a latent construct be congeneric (i.e., believed to measure the same construct), they need not exhibit invariant variances and error/uniquenesses.

When the multiple indicators of a CFA model represent items from a single measuring instrument, however, it may be of interest to test for the invariance of item reliabilities as a means to detect evidence of item bias (see e.g., Benson, 1987). In contrast to the conceptual definition of item bias generally associated with cognitive instruments (i.e., individuals of equal ability have unequal probability of success), item bias related to affective instruments reflects on its validity, and hence, on the question of whether items generate the same meaning across groups; evidence of such item bias is a clear indication that the scores are differentially valid (Green, 1975).

From classical test theory, item reliability is defined as the ratio of true score variance to total score variance (true plus error score variance); in LISREL lexicon this can be represented as $\phi/(\phi + \theta_\delta)$, where ϕ represents factor true score variance and θ_δ represents error score variance associated with measures of the factor. For example, in the present study, $\phi_{11}/(\phi_{11} + \theta_{\delta 11})$ represents the ratio of true score to total score variance for general SC. (Total score variance = true score variance plus error score variance associated with the six item-pair measurements of general SC.) It follows from this that the reliability of each measure in a LISREL model is determined in part by the variance of its corresponding factor; the reliability ratio thus becomes $\lambda^2\phi/(\lambda^2\phi + \theta_\delta)$ (Jöreskog, 1971b). Again, within the framework of the present data, $\lambda_{31}^2\phi_{11}/(\lambda_{31}^2\phi_{11} + \theta_{\delta 33})$ would represent the reliability of the third item-pair designed to measure general SC.

In examining test reliability, it is important to know if the factor variances are equivalent across groups. If they are, then the invariance of item reliabilities is tested by constraining related λs, δs, and ϕs across groups (see e.g., Cole & Maxwell, 1985; Rock et al., 1978). If, on the other hand, the factor variances are nonequivalent across groups, then testing for reliability invariance must be based on the ratio of true and

TABLE 7.4. LISREL Specification Input for Model 4 (High Track Only): Testing for Invariant Factor Variances and Covariances

```
Testing for Invariance of Item Pairs - Grp 2 = Hi Track
DA NO=613
LA
*
'GSC1' 'GSC2' 'GSC3' 'GSC4' 'GSC5' 'GSC6' 'ASC1' 'ASC2' 'ASC3' 'ASC4'
'ASC5' 'ESC1' 'ESC2' 'ESC3' 'ESC4' 'ESC5' 'MSC1' 'MSC2' 'MSC3' 'MSC4'
'MSC5'
KM SY
(21F3.2)
1000
 71100
 79 69100
 63 57 62100
 74 71 77 69100
 62 59 61 68 64100
 25 24 19 28 19 24100
 21 22 18 30 26 23 70100
 23 22 23 31 21 24 55 78100
 25 22 24 33 23 24 50 70 76100
 25 23 21 31 21 29 42 61 72 69100
 15 04 14 09 10 08 14 22 24 26 23100
 13 11 13 14 14 07 18 29 28 33 23 41100
 25 18 31 21 25 22 12 24 30 35 28 42 44100
 33 24 32 21 27 23 19 27 30 42 32 45 47 42100
 04 01 03 10 08 05 08 19 19 21 21 19 35 24 24100
 14 11 09 17 12 14 34 39 34 36 36-06 02-02 08 11100
 11 12 12 13 11 14 18 22 25 26 35-15-06-08 02 02 71100
 11 11 11 10 11 15 21 32 35 35 42-10 01-00 06 08 75 81100
 12 11 11 15 08 15 25 34 39 40 47-05 05 05 11 08 69 78 82100
 08 07 07 15 09 13 28 33 32 34 41-06 04 02 10 05 67 72 72 74100
ME
*
6.21 6.40 5.96 6.47 6.55 6.17 4.66 5.96 5.90 6.02 6.17 5.52
5.81 5.92 5.50 7.10 5.30 4.64 5.29 5.19 4.13
SD
*
1.38 1.30 1.39 1.36 1.48 1.39 1.58 1.33 1.38 1.30 1.40 1.71
1.62 1.31 1.43 1.52 1.88 2.05 1.86 1.78 1.82
MO LX=IN PH=IN TD=SY,FI
FR TD(1,1)  TD(2,2)  TD(3,3)  TD(4,4)  TD(5,5)  TD(6,6)  TD(7,7)  TD(8,8)
FR TD(9,9)  TD(10,10) TD(11,11) TD(12,12) TD(13,13) TD(14,14) TD(15,15)
FR TD(16,16) TD(17,17) TD(18,18) TD(19,19 TD(20,20) TD(21,21)
ST .2 TD(1,1)  TD(2,2)  TD(3,3)  TD(4,4)  TD(5,5)  TD(6,6)  TD(7,7)  TD(8,8)
ST .2 TD(9,9)  TD(10,10) TD(11,11) TD(12,12) TD(13,13) TD(14,14) TD(15,15)
ST .2 TD(16,16) TD(17,17) TD(18,18) TD(19,19 TD(20,20) TD(21,21)
OU NS
```

error score variances (Cole & Maxwell, 1985; Rock et al., 1978). This procedure, however, is quite complex and has not been fully demonstrated in the literature. Although Werts et al. (1976) address the testing of a ratio of variances, they do not provide an explicit application related to tests for the invariance of the reliability ratio.

Returning once again to our track data, our first step is to test for the invariance of factor variances in order to establish the viability of imposing equality constraints on the λ and δ for each item-pair. The LISREL input for this model (Model 4), as specified for the high track, is presented in Table 7.4.[4]

---

[4]Although this model was specified by adding separate equality constraint statements for each of the variance parameters PH(1,1), PH(2,2), PH(3,3), and PH(4,4), we could just as easily have specified the entire Φ matrix invariant (PH = IN) on the MO card.

TABLE 7.5. Summary of Invariance Tests for Item-Pair Measurements and Structure Across Track

Competing models	χ^2	df	$\Delta\chi^2$	Δdf
1. Four SC factors invariant	1,230.64	366	—	—
2. Model 1 with pattern of factor loadings invariant	1,254.55	383	23.91	17
3. Model 2 with factor covariances invariant	1,260.97	389	6.42	6
4. Model 3 with factor variances and covariances constrained invariant	1,288.58	393	27.61***	4
5. Model 3 with factor variances independently invariant				
a) General SC	1,265.02	390	4.05*	1
b) Academic SC	1,274.10	390	13.13***	1
c) English SC	1,261.05	390	.08	1
d) Mathematics SC	1,275.76	390	14.79***	1

*p<0.05   ***p<0.001
SC = self-concept.

The difference in model fit between Models 3 and 4 was found to be highly significant ($\Delta\chi^2$ (4) = 27.61, $p<0.001$); the hypothesis of equivalent factor variances must therefore be rejected. Given these findings, we now want to determine which of the four variances are noninvariant in order that we know how to proceed in testing for the invariance of item-pair reliabilities. Thus, we proceed to test for the equality of each variance parameter, independently, using the same procedure as that demonstrated in Chapter 6. As such, Model 3 (in which all factor loadings and covariances are constrained equal) is estimated with the additional specification that the variance of general SC (ϕ_{11}) be constrained equal across track; likewise, each of the remaining variances for academic (ϕ_{22}), English (ϕ_{33}), and mathematics (ϕ_{44}) SCs is specified, respectively, and the model subsequently estimated. For example, the LISREL input for the first of these four models tests for the equality of general SC variance; the specification is as specified in Table 7.3, but with the added statement—EQ PH(1,1,1) PH(1,1).

Results of these tests were derived from the comparison of each of these models with Model 3, in which only the factor loadings and covariances were constrained. Findings revealed only one factor variance (ϕ_{33}, English SC) to be invariant across track ($\Delta\chi^2_{(1)}$ = 0.08). Results for the three noninvariant variances are as follows: general SC ($\Delta\chi^2_{(1)}$ = 4.05, $p<0.05$); academic SC ($\Delta\chi^2_{(1)}$ = 13.13, $p<0.001$); and mathematics SC ($\Delta\chi^2_{(1)}$ = 14.79, $p<0.001$). These results are summarized in Table 7.5.

For purposes of the present study, further tests of invariance are conducted for the English SC factor only. Our first concern, then, is to determine if all item-pairs comprising this subscale are invariant across track. We begin by testing a model in which all the factor loadings and covari-

TABLE 7.6. Summary of Invariance Tests for Item-Pair Reliabilities Across Track[a]

Competing models	χ^2	df	$\Delta\chi^2$	Δdf
3. All factor loadings and covariances invariant	1,260.97	389	—	—
Model 3 with:				
a) ESC subscale error variances ($\delta_{12.12}-\delta_{16.16}$) invariant	1,281.91	394	20.94[***]	5
b) $\delta_{12.12}$ invariant	1,261.44	390	.47	1
c) $\delta_{12.12}$; $\delta_{13.13}$ invariant	1,261.45	391	.48	2
d) $\delta_{12.12}$; $\delta_{13.13}$ $\delta_{14.14}$ invariant	1,267.32	392	6.35	3
e) $\delta_{12.12}$; $\delta_{13.13}$; $\delta_{14.14}$; $\delta_{15.15}$ invariant	1,270.54	393	9.57[*]	4
f) $\delta_{12.12}$; $\delta_{13.13}$; $\delta_{14.14}$; $\delta_{16.16}$ invariant	1,278.55	393	12.58[**]	4

[*]$p<0.05$ [**]$p<0.01$ [***]$p<0.001$
[a]English self-concept subscale only.
ESC = English self-concept.

ances are constrained equal across groups (Model 3), but with the added constraint that the error variance for each item-pair measuring English SC ($\delta_{12.12}, \delta_{13.13}, \delta_{14.14}, \delta_{15.15}, \delta_{16.16}$) is also held invariant; Model 3 serves as the base model against which all competing models related to the reliability of English SC measures are compared. The additional LISREL input for testing the entire subscale would therefore specify the following:

$$\text{EQ TD}(1,12,12) \text{ TD}(12,12).$$
$$\text{ED TD}(1,13,13) \text{ TD}(13,13).$$
$$\text{ED TD}(1,14,14) \text{ TD}(14,14).$$
$$\text{ED TD}(1,15,15) \text{ TD}(15,15).$$
$$\text{ED TD}(1,16,16) \text{ TD}(16,16).$$

Results from these tests are summarized in Table 7.6. As shown here, the test of equal reliability of all English SC measures in combination demonstrated a significant $\Delta\chi^2$ ($p<0.001$) indicating that the reliability of at least one item-pair was noninvariant across track. Given these findings, as illustrated in Chapter 6, each item-pair measurement was subsequently tested independently in order to detect the noninvariant measures; item-pairs 15 and 16 were found to be noninvariant across track.

3. Summary

This chapter examined tests for the invariance of a single measuring instrument across levels of academic ability. Specifically, hypotheses were tested that related to the equivalence of the SDQIII subscale item measurements, theoretical relations among the four facets of SC (general, academic, English, and mathematics), and reliabilities of the item-pairs related to each subscale. Tests for invariance were conducted at both the matrix and individual parameter levels.

8
Testing for Invariant Latent Mean Structures

The primary purpose of this chapter is to demonstrate procedures for testing the invariance of latent mean structures, or, stated differently, testing for differences in latent means. The analytic strategy here differs both conceptually and technically from the two previous applications related to invariance. A secondary aim of this chapter is to demonstrate the technique of testing for and with partial measurement invariance; thus far, we have not had occasion to apply this important procedure. (For details of the study related to this application, see Byrne, 1988a.)

Let us first review the basic conceptual and technical differences in tests of invariance related to the current and former applications. In Chapters 6 and 7 our analyses involved testing for the invariance of factor measurement and variance-covariance parameters; as such, only the analysis of covariance structures was of interest. This is because in such analyses, modeling with the mean-related parameters does not impose restrictions on the observed variable means. In testing for the invariance of factor means, on the other hand, the modeling does involve restrictions on the observed variable means and, therefore, the analysis is based on both the covariance and mean structures. Thus, in addition to the factor measurement (Λ, Θ) and factor variance-covariance matrices (Φ), the regression intercept (nu, ν) and mean (gamma, Γ) vectors are of primary interest. More specifically, ν is a vector of constant intercept terms and considered to be a component of the LISREL measurement model; Γ is a vector of mean estimates and a component of the LISREL structural model.

Technically, testing for the invariance of mean structures is more complicated and thus more tedious than testing for the invariance of covariance structures. (The paucity of reported research in which the technique has been applied would seem to attest to this fact; for a review, see Byrne et al., in press.) For example, in order to test for the invariance of latent means, the model must be structured as an all-Y specification. This means that if the researcher has used an all-X model in preceding analyses (as we have done thus far in this book), a reformulation of model specifications is a necessary prerequisite to further analyses involving mean structures. The primary aim of the present chapter is to outline, in some detail, the steps involved in transforming an all-X to an all-Y model in order to test for latent mean differences (i.e., the invariance of latent mean structures).

1. Tests for Invariance Related to Latent Self-Concept Means

1.1. The Hypothesized Model

The postulated model in the present application is identical to the one proposed in Chapter 6 (see Figure 6.1), with one exception—the present model hypothesizes that SC measurements and structure are factorially invariant across low- and high-academic tracks (rather than across gender).

1.2. The Baseline Models

For both the low- and high-track groups, the baseline model was derived from the process of post hoc model fitting with concomitant sensitivity analyses. These investigations led to final models that included one secondary loading (low track = λ_{71}; high track = λ_{61}), and four error covariances between subscales of the same measuring instrument, three of which were common across track. (For a more extensive discussion of analyses related to the fitting of these models, see Byrne et al., 1989.) A summary of model specifications related to fitting the baseline models for low and high track is presented in Table 8.1; the models are illustrated schematically in Figure 8.1.

2. Testing for the Invariance of Factor Covariance Structures

As with previous tests for invariance demonstrated in Chapters 6 and 7, the simultaneous estimation of parameters for both tracks was based on covariance, rather than on correlation matrices. One marked difference between this application and those presented earlier is that the baseline model, for both the low and high-tracks, includes one secondary loading[1] that is dissimilar across track. Thus, it is important to note that these

---

[1]Secondary loadings are measurement loadings on more than one factor; they are also referred to as cross-loadings.

TABLE 8.1. Steps in Fitting the Baseline Model

Competing models	χ^2	df	$\Delta\chi^2$	df	χ^2/df	BBI	TLI
			Low Track				
0 Null model	1,429.60	55	—	—	25.99	—	—
1 Basic four-factor Model with $\theta_i\theta_j = 0$	160.54	38	—	—	4.22	.89	.87
2 $\theta_{10.7}$, free	122.24	37	38.30***	1	3.30	.91	.91
3 $\theta_{10.7}$, θ_{85} free	97.95	36	24.29***	1	2.72	.93	.93
4 $\theta_{10.7.}$, θ_{85}, $\theta_{11.5}$ free	71.38	35	26.57***	1	2.04	.95	.96
5 $\theta_{10.7}$, θ_{85}, $\theta_{11.5}$ free λ_{71} free	54.80	34	16.58***	1	1.61	.96	.98
6 $\theta_{10.7}$, θ_{85}, $\theta_{11.5}$, θ_{96}, free λ_{71} free	49.10	33	5.70*	1	1.49	.97	.98
			High Track				
0 Null model	4,784.85	55			87.00	—	—
1 Basic four-factor Model with $\theta_i\theta_j = 0$	401.09	38	—	—	10.56	.92	.89
2 θ_{85} free	277.67	37	123.42***	1	7.50	.94	.92
3 θ_{85}, $\theta_{11.5}$ free	192.50	36	85.17***	1	5.35	.96	.95
4 θ_{85}, $\theta_{11.5}$, $\theta_{10.7}$, free	153.91	35	38.59***	1	4.40	.97	.96
5 θ_{85}, $\theta_{11.5}$, $\theta_{10.7}$ free λ_{61} free	126.86	34	27.05***	1	3.73	.97	.97
6 θ_{85}, $\theta_{11.5}$, $\theta_{10.7}$, $\theta_{11.8}$ free λ_{61} free	105.60	33	21.26***	1	3.20	.98	.97

*$p < 0.05$ ***$p < 0.001$

From Byrne, Shavelson, and Muthén (1989) "Testing for the Equivalence of Factor Covariance and Mean Structures: The Issue of Partial Measurement Invariance" in *Psychological Bulletin, 105*, 456–466. Copyright 1989 by American Psychological Association. Reprinted with permission.

secondary loading parameters must remain unconstrained throughout the invariance testing procedures. The LISREL specification input for the simultaneous estimation of these baseline models is presented in Table 8.2.[2]

2.1. Testing for the Invariance of Measurement Parameters

Invariant Number of Factors. Consistent with our previous invariance testing procedures, the hypothesis of an invariant number of factors was tested first (Model 1). This simultaneous four-factor solution yielded a substantively reasonable fit to the data (BBI = 0.98; TLI = 0.99)[3] sug-

---

[2]Note the reverse order of the group input data here compared with that in Chapter 6 (i.e., group 1 = high track; group 2 = low track).

[3]The TLI (Tucker-Lewis index), like the BBI, is indicative of the percentage of covariance explained by the hypothesized model; a value <0.90 usually means that the model can be improved substantially.

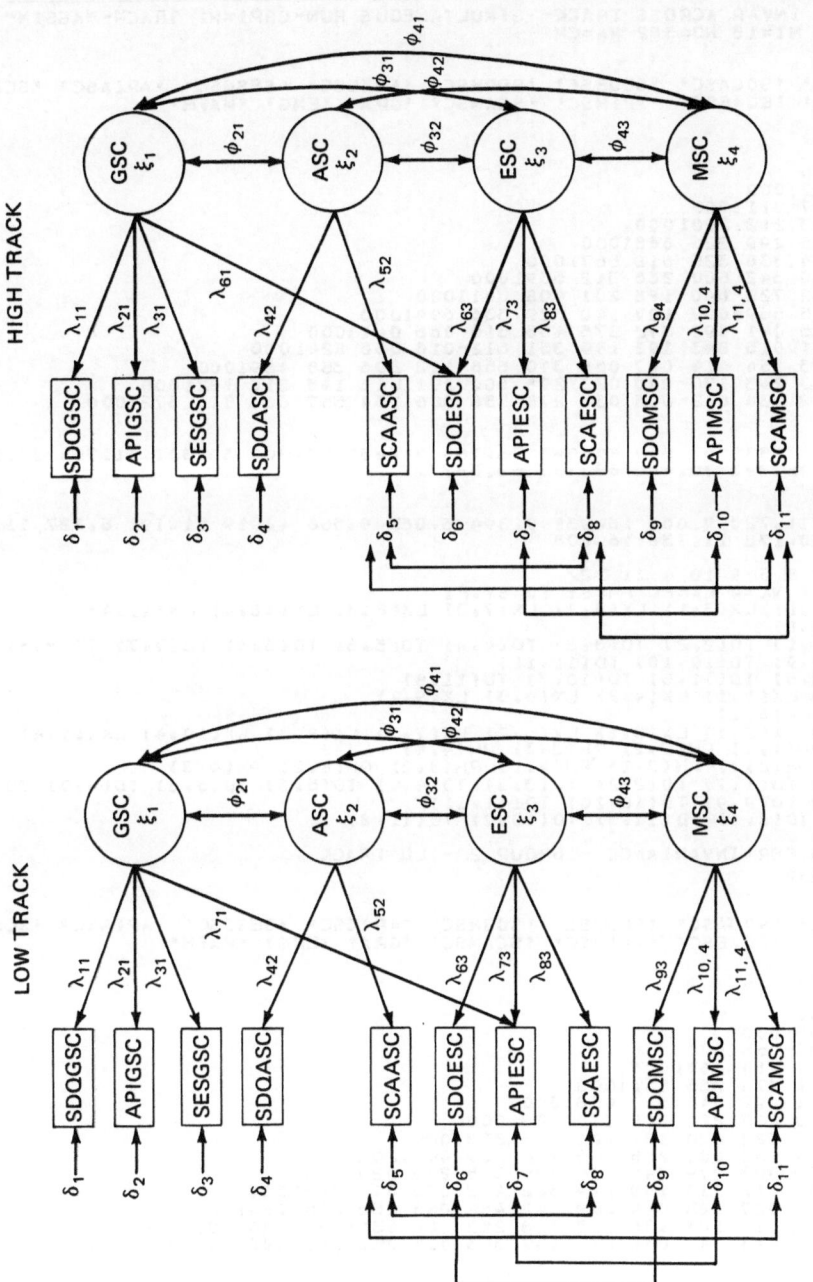

FIGURE 8.1. Baseline Models of Self-Concept Structure for Low and High Tracks.

TABLE 8.2. LISREL Specification Input for the Simultaneous Estimation of
Low- and High-Track Baseline Models

```
TESTING INVAR ACROSS TRACK- SIMULTANEOUS RUN-GRP1=HI TRACK-"AGSIM" FILE
DA NG=2 NI=15 NO=582 MA=CM
LA
*
'SDQGSC' 'SDQASC' 'SDQESC' 'SDQMSC' 'APIGSC' 'SESGSC' 'APIASC' 'SCAASC'
'APIESC' 'SCAESC' 'APIMSC' 'SCAMSC' 'GPA' 'ENG' 'MATH'
KM SY
(15F4.3)
1000
 3301000
 267 3971000
 173 427-0111000
 658 323 212 2001000
 812 325 290 226 6681000
 556 624 338 325 618 5671000
 250 660 342 500 266 312 5391000
 151 412 723-040 188 201 408 3411000
 100 415 559-007 107 140 329 536 6941000
 180 455 041 892 272 275 405 516 066 0411000
 133 401 015 843 193 189 351 612-016 086 8241000
 016 493 154 414 007 065 340 658 112 296 358 4891000
-028 423 243 170-070 023 278 500 291 516 145 218 7821000
-015 372 038 612-005 036 278 558 006 134 557 666 811 5721000
ME
*
75.792 57.330 57.569 49.043 76.768 31.467 73.802 30.301 61.794 28.933
47.223 26.223 70.440 68.787 62.687
SD
*
14.563 11.723 9.867 16.951 9.394 5.063 9.556 4.919 11.191 5.727 11.606
7.986 10.172 11.738 16.208
SE
1 5 6 2 8 3 9 10 4 11 12/
MO NX=11 NK=4 LX=FU PH=SY TD=SY,FI
FR LX(2,1) LX(3,1) LX(5,2) LX(7,3) LX(8,3) LX(10,4) LX(11,4)
FR LX(6,1)
FR TD(1,1) TD(2,2) TD(3,3) TD(4,4) TD(5,5) TD(6,6) TD(7,7) TD(8,8)
FR TD(9,9) TD(10,10) TD(11,11)
FR TD(8,5) TD(11,5) TD(10,7) TD(11,8)
ST 30.0 LX(1,1) LX(4,2) LX(6,3) LX(9,4)
ST 5.0 LX(6,1)
ST 15.0 LX(2,1) LX(3,1) LX(5,2) LX(7,3) LX(8,3) LX(10,4) LX(11,4)
ST .1 PH(1,1) PH(2,2) PH(3,3) PH(4,4)
ST .05 PH(2,1) PH(3,1) PH(4,1) PH(3,2) PH(4,2) PH(4,3)
ST 40.0 TD(1,1) TD(2,2) TD(3,3) TD(4,4) TD(5,5) TD(6,6) TD(7,7) TD(8,8)
ST 40.0 TD(9,9) TD(10,10) TD(11,11)
ST 6.0 TD(8,5) TD(11,5) TD(10,7) TD(11,8)
OU NS
TESTING FOR INVARIANCE - GROUP 2 - LO TRACK
DA NO=248
LA
*
'SDQGSC' 'SDQASC' 'SDQESC' 'SDQMSC' 'APIGSC' 'SESGSC' 'APIASC' 'SCAASC'
'APIESC' 'SCAESC' 'APIMSC' 'SCAMSC' 'GPA' 'ENG' 'MATH'
KM SY
(15F4.3)
1000
 3201000
 307 2981000
 244 355-0551000
 614 237 214 2671000
 755 261 276 255 5881000
 456 571 392 345 547 4581000
 270 580 265 226 219 270 5231000
 143 430 523 030 181 108 476 3731000
 231 377 433 004 265 245 424 509 4981000
 250 388 063 779 245 234 409 345 232 0271000
 234 348-012 719 199 214 362 442 075 077 7421000
 038 361 027 125 049 009 342 452 063 256 035 1541000
 059 320 073 103 135 038 318 252 144 435-020 010 6971000
 009 232 014 344 036-021 199 355 065 086 280 490 647 3421000
ME
*
75.936 49.415 55.036 41.569 76.758 31.157 70.165 24.746 57.794 25.343
41.710 22.944 61.149 58.875 59.391
SD
```

TABLE 8.2. Continued

```
#
13.442 12.391 9.468 13.416 9.028 4.875 8.830 4.480 10.701 4.858 10.566
5.824 9.149 11.857 14.854
SE
1 5 6 2 8 3 9 10 4 11 12/
MO LX=FU PH=SY TD=SY,FI
FR LX(2,1) LX(3,1) LX(5,2) LX(7,3) LX(8,3) LX(10,4) LX(11,4)
FR LX(7,1)
FR TD(1,1) TD(2,2) TD(3,3) TD(4,4) TD(5,5) TD(6,6) TD(7,7) TD(8,8)
FR TD(9,9) TD(10,10) TD(11,11)
FR TD(8,5) TD(11,5) TD(10,7) TD(9,6)
ST 30.0 LX(1,1) LX(4,2) LX(6,3) LX(9,4)
ST 15.0 LX(2,1) LX(3,1) LX(5,2) LX(7,3) LX(8,3) LX(10,4) LX(11,4)
ST 5.0 LX(7,1)
ST .1  PH(1,1) PH(2,2) PH(3,3) PH(4,4)
ST .05 PH(2,1) PH(3,1) PH(4,1) PH(3,2) PH(4,2) PH(4,3)
ST 40.0 TD(1,1) TD(2,2) TD(3,3) TD(4,4) TD(5,5) TD(6,6) TD(7,7) TD(8,8)
ST 40.0 TD(9,9) TD(10,10) TD(11,11)
ST 6.0 TD(8,5) TD(11,5) TD(10,7) TD(9,6)
OU NS
```

gesting that, for both tracks, the data were well described by four SC factors—general SC, academic SC, English SC, and mathematics SC.

Invariant Pattern of Factors Loadings. The hypothesis of an invariant pattern of factor loadings was tested next by constraining all lambda parameters (except λ_{71} and λ_{61}) to be equal, and then comparing this model (Model 2) with Model 1, in which the number of factors and pattern of loadings were held invariant across track, but not constrained equal. Given the two different cross-loadings (i.e., secondary loadings) for each track, the entire Λ matrix could not be constrained equal on the MO card using the specification LX = IN. Rather, a separate statement of equality had to be specified for each λ parameter. This LISREL specification input for the low track only (as Group 2) is presented in Table 8.3.

TABLE 8.3. LISREL Specification Input for Model 2 (Low Track Only): Testing for an Invariant Pattern of Factor Loadings

```
MO LX=FU PH=SY TD=SY,FI
FR LX(2,1) LX(3,1) LX(5,2) LX(7,3) LX(8,3) LX(10,4) LX(11,4)
FR LX(7,1)
FR TD(1,1) TD(2,2) TD(3,3) TD(4,4) TD(5,5) TD(6,6) TD(7,7) TD(8,8)
FR TD(9,9) TD(10,10) TD(11,11)
FR TD(8,5) TD(11,5) TD(10,7) TD(9,6)
ST 30.0 LX(1,1) LX(4,2) LX(6,3) LX(9,4)
ST 15.0 LX(7,3) LX(8,3) LX(10,4) LX(11,4)
ST 5.0 LX(7,1)
ST .1  PH(1,1) PH(2,2) PH(3,3) PH(4,4)
ST .05 PH(2,1) PH(3,1) PH(4,1) PH(3,2) PH(4,2) PH(4,3)
ST 40.0 TD(1,1) TD(2,2) TD(3,3) TD(4,4) TD(5,5) TD(6,6) TD(7,7) TD(8,8)
ST 40.0 TD(9,9) TD(10,10) TD(11,11)
ST 6.0 TD(8,5) TD(11,5) TD(10,7) TD(9,6)
EQ LX(1,2,1) LX(2,1)
EQ LX(1,3,1) LX(3,1)
EQ LX(1,5,2) LX(5,2)
EQ LX(1,7,3) LX(7,3)
EQ LX(1,8,3) LX(8,3)
EQ LX(1,10,4) LX(10,4)
EQ LX(1,11,4) LX(11,4)
OU NS
```

The estimation of this model yielded a χ^2 differential that was highly significant ($\Delta\chi^2_{(7)} = 25.82$, $p < 0.001$), indicating that the hypothesis of an invariant pattern of factor loadings must be rejected. Confronted with this finding, it is now of interest to pinpoint differences in the measurement parameters between low and high tracks. As such, we test for partial measurement invariance by testing, independently, the equivalence of each congeneric set of lambda parameters specified to measure each SC facet.

2.2. Testing for Partial Measurement Invariance

We begin by examining the measurement of general SC, by holding λ_{21} and λ_{31} invariant across track. This hypothesized model is found to be tenable ($\Delta\chi^2$ (2) = 0.16). Thus, we next test the equality of measurements for academic SC by holding λ_{21}, λ_{31}, and λ_{52} invariant; this hypothesized model could also not be rejected ($\Delta\chi^2$ (3) = 7.41). Continuing in this manner, we test the invariance of factor measurements for both the English and mathematics SC factors; both hypothesized models were rejected (English SC, $\Delta\chi^2$ (5) = 17.01, $p<0.01$; mathematics SC, $\Delta\chi^2$ (5) = 16.59, $p<0.01$). Before proceeding further with these tests of partial measurement invariance, it seems prudent to stop and allow the reader to review the LISREL specification input for each of these tests. This information is presented in Table 8.4. In examining these model specifications, it is important to note the cumulative pattern of adding invariant parameters to each successive model to be estimated.

As we noted earlier the combination of instruments measuring English and mathematics SCs was found to be noninvariant across track. Now we want to determine if any one of these measures is actually invariant across track. Thus, we proceed to test for the equality of each of these lambda parameters, individually, while concomitantly holding λ_{21}, λ_{31}, and λ_{52} invariant. You can study these model specifications in Table 8.5.

These results revealed that English SC as measured by the SCA (λ_{83}), and mathematics SC as measured by the API ($\lambda_{10,4}$), were inconsistent across track. In other words, the SCA and API were not measuring English SC and mathematics SC, respectively, in the same way for students in both academic tracks. A summary of findings related to tests for the equality of the SC measuring instruments across track is detailed in Table 8.6.

Admittedly, the sequential testing of models in the exploration of partial measurement invariance is problematic. Given the nonindependence of the tests, it is possible that an alternative series of tests might lead to quite different results. While we might believe that our sequential model-fitting procedures are substantively reasonable, verification must come from cross-validated studies.

TABLE 8.4. LISREL Specification Input for Models 3–6 (Low Track Only): Testing for Invariant Subscales

Model 3

```
MO LX=FU PH=SY TD=SY,FI
FR LX(2,1) LX(3,1) LX(5,2) LX(7,3) LX(8,3) LX(10,4) LX(11,4)
FR LX(7,1)
FR TD(1,1) TD(2,2) TD(3,3) TD(4,4) TD(5,5) TD(6,6) TD(7,7) TD(8,8)
FR TD(9,9) TD(10,10) TD(11,11)
FR TD(8,5) TD(11,5) TD(10,7) TD(9,6)
ST 30.0 LX(1,1) LX(4,2) LX(6,3) LX(9,4)
ST 15.0 LX(7,3) LX(8,3) LX(10,4) LX(11,4)
ST 5.0 LX(7,1)
ST .1  PH(1,1) PH(2,2) PH(3,3) PH(4,4)
ST .05 PH(2,1) PH(3,1) PH(4,1) PH(3,2) PH(4,2) PH(4,3)
ST 40.0 TD(1,1) TD(2,2) TD(3,3) TD(4,4) TD(5,5) TD(6,6) TD(7,7) TD(8,8)
ST 40.0 TD(9,9) TD(10,10) TD(11,11)
ST 6.0 TD(8,5) TD(11,5) TD(10,7) TD(9,6)
EQ LX(1,2,1) LX(2,1)
EQ LX(1,3,1) LX(3,1)
OU NS
```

Model 4

```
MO LX=FU PH=SY TD=SY,FI
FR LX(2,1) LX(3,1) LX(5,2) LX(7,3) LX(8,3) LX(10,4) LX(11,4)
FR LX(7,1)
FR TD(1,1) TD(2,2) TD(3,3) TD(4,4) TD(5,5) TD(6,6) TD(7,7) TD(8,8)
FR TD(9,9) TD(10,10) TD(11,11)
FR TD(8,5) TD(11,5) TD(10,7) TD(9,6)
ST 30.0 LX(1,1) LX(4,2) LX(6,3) LX(9,4)
ST 15.0 LX(7,3) LX(8,3) LX(10,4) LX(11,4)
ST 5.0 LX(7,1)
ST .1  PH(1,1) PH(2,2) PH(3,3) PH(4,4)
ST .05 PH(2,1) PH(3,1) PH(4,1) PH(3,2) PH(4,2) PH(4,3)
ST 40.0 TD(1,1) TD(2,2) TD(3,3) TD(4,4) TD(5,5) TD(6,6) TD(7,7) TD(8,8)
ST 40.0 TD(9,9) TD(10,10) TD(11,11)
ST 6.0 TD(8,5) TD(11,5) TD(10,7) TD(9,6)
EQ LX(1,2,1) LX(2,1)
EQ LX(1,3,1) LX(3,1)
EQ LX(1,5,2) LX(5,2)
OU NS
```

Model 5

```
MO LX=FU PH=SY TD=SY,FI
FR LX(2,1) LX(3,1) LX(5,2) LX(7,3) LX(8,3) LX(10,4) LX(11,4)
FR LX(7,1)
FR TD(1,1) TD(2,2) TD(3,3) TD(4,4) TD(5,5) TD(6,6) TD(7,7) TD(8,8)
FR TD(9,9) TD(10,10) TD(11,11)
FR TD(8,5) TD(11,5) TD(10,7) TD(9,6)
ST 30.0 LX(1,1) LX(4,2) LX(6,3) LX(9,4)
ST 15.0 LX(7,3) LX(8,3) LX(10,4) LX(11,4)
ST 5.0 LX(7,1)
ST .1  PH(1,1) PH(2,2) PH(3,3) PH(4,4)
ST .05 PH(2,1) PH(3,1) PH(4,1) PH(3,2) PH(4,2) PH(4,3)
ST 40.0 TD(1,1) TD(2,2) TD(3,3) TD(4,4) TD(5,5) TD(6,6) TD(7,7) TD(8,8)
ST 40.0 TD(9,9) TD(10,10) TD(11,11)
ST 6.0 TD(8,5) TD(11,5) TD(10,7) TD(9,6)
EQ LX(1,2,1) LX(2,1)
EQ LX(1,3,1) LX(3,1)
EQ LX(1,5,2) LX(5,2)
EQ LX(1,7,3) LX(7,3)
EQ LX(1,8,3) LX(8,3)
OU NS
```

Model 6

```
MO LX=FU PH=SY TD=SY,FI
FR LX(2,1) LX(3,1) LX(5,2) LX(7,3) LX(8,3) LX(10,4) LX(11,4)
FR LX(7,1)
FR TD(1,1) TD(2,2) TD(3,3) TD(4,4) TD(5,5) TD(6,6) TD(7,7) TD(8,8)
FR TD(9,9) TD(10,10) TD(11,11)
FR TD(8,5) TD(11,5) TD(10,7) TD(9,6)
ST 30.0 LX(1,1) LX(4,2) LX(6,3) LX(9,4)
ST 15.0 LX(7,3) LX(8,3) LX(10,4) LX(11,4)
ST 5.0 LX(7,1)
ST .1  PH(1,1) PH(2,2) PH(3,3) PH(4,4)
ST .05 PH(2,1) PH(3,1) PH(4,1) PH(3,2) PH(4,2) PH(4,3)
ST 40.0 TD(1,1) TD(2,2) TD(3,3) TD(4,4) TD(5,5) TD(6,6) TD(7,7) TD(8,8)
ST 40.0 TD(9,9) TD(10,10) TD(11,11)
ST 6.0 TD(8,5) TD(11,5) TD(10,7) TD(9,6)
EQ LX(1,2,1) LX(2,1)
EQ LX(1,3,1) LX(3,1)
EQ LX(1,5,2) LX(5,2)
EQ LX(1,10,4) LX(10,4)
EQ LX(1,11,4) LX(11,4)
OU NS
```

TABLE 8.5. LISREL Specification Input for Models 7–10 (Low Track Only): Testing for Partial Measurement Invariance

Model 7

```
MO LX=FU PH=SY TD=SY,FI
FR LX(2,1)  LX(3,1) LX(5,2) LX(7,3) LX(8,3) LX(10,4) LX(11,4)
FR LX(7,1)
FR TD(1,1)  TD(2,2)  TD(3,3)  TD(4,4)  TD(5,5)  TD(6,6)  TD(7,7)  TD(8,8)
FR TD(9,9)  TD(10,10)  TD(11,11)
FR TD(8,5)  TD(11,5)  TD(10,7)  TD(9,6)
ST 30.0 LX(1,1)  LX(4,2)  LX(6,3)  LX(9,4)
ST 15.0 LX(7,3)  LX(8,3)  LX(10,4)  LX(11,4)
ST 5.0 LX(7,1)
ST .1   PH(1,1)  PH(2,2)  PH(3,3)  PH(4,4)
ST .05 PH(2,1)  PH(3,1)  PH(4,1)  PH(3,2)  PH(4,2)  PH(4,3)
ST 40.0 TD(1,1)  TD(2,2)  TD(3,3)  TD(4,4)  TD(5,5)  TD(6,6)  TD(7,7)  TD(8,8)
ST 40.0 TD(9,9)  TD(10,10)  TD(11,11)
ST 6.0 TD(8,5)  TD(11,5)  TD(10,7)  TD(9,6)
EQ LX(1,2,1)  LX(2,1)
EQ LX(1,3,1)  LX(3,1)
EQ LX(1,5,2)  LX(5,2)
EQ LX(1,7,3)  LX(7,3)
OU NS
```

Model 8

```
MO LX=FU PH=SY TD=SY,FI
FR LX(2,1)  LX(3,1) LX(5,2) LX(7,3) LX(8,3) LX(10,4) LX(11,4)
FR LX(7,1)
FR TD(1,1)  TD(2,2)  TD(3,3)  TD(4,4)  TD(5,5)  TD(6,6)  TD(7,7)  TD(8,8)
FR TD(9,9)  TD(10,10)  TD(11,11)
FR TD(8,5)  TD(11,5)  TD(10,7)  TD(9,6)
ST 30.0 LX(1,1)  LX(4,2)  LX(6,3)  LX(9,4)
ST 15.0 LX(7,3)  LX(8,3)  LX(10,4)  LX(11,4)
ST 5.0 LX(7,1)
ST .1   PH(1,1)  PH(2,2)  PH(3,3)  PH(4,4)
ST .05 PH(2,1)  PH(3,1)  PH(4,1)  PH(3,2)  PH(4,2)  PH(4,3)
ST 40.0 TD(1,1)  TD(2,2)  TD(3,3)  TD(4,4)  TD(5,5)  TD(6,6)  TD(7,7)  TD(8,8)
ST 40.0 TD(9,9)  TD(10,10)  TD(11,11)
ST 6.0 TD(8,5)  TD(11,5)  TD(10,7)  TD(9,6)
EQ LX(1,2,1)  LX(2,1)
EQ LX(1,3,1)  LX(3,1)
EQ LX(1,5,2)  LX(5,2)
EQ LX(1,8,3)  LX(8,3)
OU NS
```

Model 9

```
MO LX=FU PH=SY TD=SY,FI
FR LX(2,1)  LX(3,1) LX(5,2) LX(7,3) LX(8,3) LX(10,4) LX(11,4)
FR LX(7,1)
FR TD(1,1)  TD(2,2)  TD(3,3)  TD(4,4)  TD(5,5)  TD(6,6)  TD(7,7)  TD(8,8)
FR TD(9,9)  TD(10,10)  TD(11,11)
FR TD(8,5)  TD(11,5)  TD(10,7)  TD(9,6)
ST 30.0 LX(1,1)  LX(4,2)  LX(6,3)  LX(9,4)
ST 15.0 LX(7,3)  LX(8,3)  LX(10,4)  LX(11,4)
ST 5.0 LX(7,1)
ST .1   PH(1,1)  PH(2,2)  PH(3,3)  PH(4,4)
ST .05 PH(2,1)  PH(3,1)  PH(4,1)  PH(3,2)  PH(4,2)  PH(4,3)
ST 40.0 TD(1,1)  TD(2,2)  TD(3,3)  TD(4,4)  TD(5,5)  TD(6,6)  TD(7,7)  TD(8,8)
ST 40.0 TD(9,9)  TD(10,10)  TD(11,11)
ST 6.0 TD(8,5)  TD(11,5)  TD(10,7)  TD(9,6)
EQ LX(1,2,1)  LX(2,1)
EQ LX(1,3,1)  LX(3,1)
EQ LX(1,5,2)  LX(5,2)
EQ LX(1,10,4)  LX(10,4)
OU NS
```

TABLE 8.5. Continued

Model 10

```
MO  LX=FU PH=SY TD=SY,FI
FR  LX(2,1) LX(3,1) LX(5,2) LX(7,3) LX(8,3) LX(10,4) LX(11,4)
FR  LX(7,1)
FR  TD(1,1) TD(2,2) TD(3,3) TD(4,4) TD(5,5) TD(6,6) TD(7,7) TD(8,8)
FR  TD(9,9) TD(10,10) TD(11,11)
FR  TD(8,5) TD(11,5) TD(10,7) TD(9,6)
ST  30.0 LX(1,1) LX(4,2) LX(6,3) LX(9,4)
ST  15.0 LX(7,3) LX(8,3) LX(10,4) LX(11,4)
ST  5.0 LX(7,1)
ST  .1   PH(1,1) PH(2,2) PH(3,3) PH(4,4)
ST  .05 PH(2,1) PH(3,1) PH(4,1) PH(3,2) PH(4,2) PH(4,3)
ST  40.0 TD(1,1) TD(2,2) TD(3,3) TD(4,4) TD(5,5) TD(6,6) TD(7,7) TD(8,8)
ST  40.0 TD(9,9) TD(10,10) TD(11,11)
ST  6.0 TD(8,5) TD(11,5) TD(10,7) TD(9,6)
EQ  LX(1,2,1) LX(2,1)
EQ  LX(1,3,1) LX(3,1)
EQ  LX(1,5,2) LX(5,2)
EQ  LX(1,11,4) LX(11,4)
OU  NS
```

TABLE 8.6. Simultaneous Tests of Invariance for Self-Concept Measurements

Competing models	χ^2	df	$\Delta\chi^2$	Δdf	χ^2/df
1 Four SC factors invariant	154.60[***]	66	—	—	2.34
2 Model 1 with major loadings on each SC factor invariant[a]	180.42	73	25.82[***]	7	2.47
3 Model 1 with major loadings on GSC invariant	154.76	68	.16	2	2.28
4 Model 1 with major loadings on GSC and ASC invariant	162.01	69	7.41	3	2.35
5 Model 1 with major loadings on GSC, ASC, and ESC invariant	171.61	71	17.01[**]	5	2.42
6 Model 1 with major loadings on GSC, ASC, and MSC invariant	171.19	71	16.59[**]	5	2.41
7 Model 4 with APIESC invariant	163.30	70	1.29	1	2.33
8 Model 4 with SCAESC invariant	166.24	70	4.23[*]	1	2.37
9 Model 4 with APIMSC invariant	169.50	70	7.49[**]	1	2.42
10 Model 4 with SCAMSC invariant	162.06	70	.05	1	2.36

[*]$p < 0.05$ [**]$p < 0.01$ [***]$p < 0.001$

[a]All lambda parameters invariant except λ_{71} and λ_{61}

GSC = general SC; ASC = academic SC; ESC = English SC; MSC = mathematics SC; APIESC = API English Perceptions subscale; SCAESC = SCA form B (SC of English ability); APIMSC = API Mathematics Perceptions subscale; SCAMSC = form C (SC of mathematics ability).

From Byrne, Shavelson, and Muthén (1989) "Testing for the Equivalence of Factor Covariance and Mean Structures: The Issue of Partial Measurement Invariance" in *Psychological Bulletin*, *105*, 456–466. Copyright 1989 by American Psychological Association. Reprinted with permission.

TABLE 8.7. LISREL Specification Input for Testing the Invariance of Self-Construct Structure (Low Track Only)

```
MO LX=FU PH=IN TD=SY,FI
FR LX(2,1) LX(3,1) LX(5,2) LX(7,3) LX(8,3) LX(10,4) LX(11,4)
FR LX(7,1)
FR TD(1,1) TD(2,2) TD(3,3) TD(4,4) TD(5,5) TD(6,6) TD(7,7) TD(8,8)
FR TD(9,9) TD(10,10) TD(11,11)
FR TD(8,5) TD(11,5) TD(10,7) TD(9,6)
ST 30.0 LX(1,1) LX(4,2) LX(6,3) LX(9,4)
ST 15.0 LX(7,3) LX(8,3) LX(10,4) LX(11,4)
ST 5.0 LX(7,1)
ST .1  PH(1,1) PH(2,2) PH(3,3) PH(4,4)
ST .05 PH(2,1) PH(3,1) PH(4,1) PH(3,2) PH(4,2) PH(4,3)
ST 40.0 TD(1,1) TD(2,2) TD(3,3) TD(4,4) TD(5,5) TD(6,6) TD(7,7) TD(8,8)
ST 40.0 TD(9,9) TD(10,10) TD(11,11)
ST 6.0 TD(8,5) TD(11,5) TD(10,7) TD(9,6)
EQ LX(1,2,1) LX(2,1)
EQ LX(1,3,1) LX(3,1)
EQ LX(1,5,2) LX(5,2)
EQ LX(1,7,3) LX(7,3)
EQ LX(1,11,4) LX(11,4)
OU NS
```

2.3. Testing for the Invariance of Structural Parameters

Having determined equalities among the measurement parameters, our interest now focuses on testing for the equality of the structural parameters; we turn, then, to the matrix of factor variances and covariances (Φ). Our first step in testing for the equality of SC structure, is to constrain the entire factor variance-covariance matrix invariant across track. The LISREL specification input for this model is shown in Table 8.7. Note that in conducting tests for the invariance of structure it is important to maintain constraints on the measurement parameters; only those measures known to be consistent in their measurements across track, however, are held invariant.

The hypothesis of an invariant factor variance-covariance matrix was found untenable ($\Delta\chi^2_{(10)} = 47.91$, $p < 0.001$). Thus, we employ the same strategy as when confronted with a noninvariant Λ matrix; we proceed to test, independently, the equivalence of each parameter in the Φ matrix; recall again that, at all times, only those measures known to be consistent in their measurements across track are held invariant (i.e., λ_{21}, λ_{31}, λ_{52}, λ_{73}, $\lambda_{11,4}$). As with our tests of measurement parameters, the LISREL input changes from a specification on the MO card (PH = IN) to a specification of equality constraints for each individual structural parameter; each one being specified, one at a time. Since this technique was specified for the individual λ parameters, it need not be detailed again here.

The results of these analyses yielded one variance (ϕ_{44}) and two covariance (ϕ_{31}, ϕ_{42}) parameters to be noninvariant across track. A summary of findings from tests for an invariant SC structure (i.e., SC variances and covariances) is presented in Table 8.8.

TABLE 8.8. Simultaneous Tests of Invariance for Self-Concept Structure

Competing models	χ^2	df	$\Delta\chi^2$	Δdf	χ^2/df
1 Invariant measurement model[a]	163.35	71	—	—	2.30
2 Model 1 with all factor variances and covariances invariant	211.26	81	47.91***	10	2.61
3 Model 1 with latent variable parameters made independently invariant					
Variances					
a) General SC	165.75	72	2.40	1	2.30
b) Academic SC	164.87	72	1.52	1	2.29
c) English SC	163.52	72	.17	1	2.27
d) Mathematics SC	190.60	72	27.25***	1	2.65
Covariances					
a) General/academic SC	164.31	72	.96	1	2.28
b) General/English SC	167.95	72	4.60*	1	2.33
c) General/mathematics SC	163.70	72	.35	1	2.27
d) Academic/English SC	163.56	72	.21	1	2.27
e) Academic/mathematics SC	175.74	72	11.74***	1	2.44
f) English/mathematics SC	166.09	72	2.74	1	2.31

*p<0.05   ***p<0.001
[a]λ_{21}, λ_{31}, λ_{52}, λ_{73}, $\lambda_{11,4}$ held invariant.

3. Testing for the Invariance of Factor Mean Structures

3.1. LISREL Input

Using LISREL to test for the invariance of mean structures requires that we make several adjustments to the baseline model specification input as presented in Table 8.2. To enable you to follow this model transformation, the specified pattern of parameters, for both the low and high tracks, is presented in Table 8.9 for the all-X model, and in Table 8.10 for the all-Y model. Note the addition of a vector of intercepts (νs) and a vector of latent mean parameters (γs). Before proceeding further, you are urged to go back and compare the LISREL input for the two baseline models (see Table 8.2) and the pattern of parameters for these same models (see Table 8.9).

As noted earlier, our first step in testing for latent mean differences was to restructure the baseline models into an all-Y specification. As such, the factor loading (Λ_x), factor variance-covariance (Φ) and error variance-covariance (Θ_δ) matrices (see Table 8.9) were converted into the Λ_Y, Ψ, and Θ_ϵ matrices, respectively; the ξs (the latent factors) were treated as ηs in the LISREL sense (see Table 8.10). Second, the program must be "tricked" into estimating the latent means. This is done by creating a dummy variable (i.e., an extra variable, 'dummy,' was added to the variable list, making a total of 12 input variables, not 11). The dummy variable was given a fixed-Y specification equal to 1.00 (i.e., its value was constrained equal to a value of 1.00). Third, to accommodate the dummy

TABLE 8.9. Pattern of LISREL Parameters for Testing the Invariance

Low Track

X		ξ_1	ξ_2	ξ_3	ξ_4
SDQGSC		1	0	0	0
APIGSC		λ_{21}	0	0	0
SESGSC		λ_{31}	0	0	0
SDQASC		0	1	0	0
SCAASC		0	λ_{52}	0	0
SDQESC	Λ_x	0	0	1	0
APIESC		λ_{71}	0	λ_{73}	0
SCAESC		0	0	λ_{83}	0
SDQMSC		0	0	0	1
APIMSC		0	0	0	$\lambda_{10.4}$
SCAMSC		0	0	0	$\lambda_{11.4}$

GSC		ϕ_{11}			
ASC	Φ	ϕ_{21}	ϕ_{22}		
ESC		ϕ_{31}	ϕ_{32}	ϕ_{33}	
MSC		ϕ_{41}	ϕ_{42}	ϕ_{43}	ϕ_{44}

X												
SDQGSC		δ_{11}										
APIGSC		0	δ_{22}									
SESGSC		0	0	δ_{33}								
SDQASC		0	0	0	δ_{44}							
SCAASC	Θ_δ	0	0	0	0	δ_{55}						
SDQESC		0	0	0	0	0	δ_{66}					
APIESC		0	0	0	0	0	0	δ_{77}				
SCAESC		0	0	0	0	δ_{85}	0	0	δ_{88}			
SDQMSC		0	0	0	0	0	δ_{96}	0	0	δ_{99}		
APIMSC		0	0	0	0	0	0	$\delta_{10.7}$	0	0	$\delta_{10.10}$	
SCAMSC		0	0	0	0	$\delta_{11.5}$	0	0	0	0	0	$\delta_{11.11}$

Λ_x = factor loading matrix; Φ = factor variance-covariance matrix; Θ_δ = error variance-covariance matrix; ξ = observed self-concept (SC) measure; $\xi_1 - \xi_4$ = SC factors (ξ_1 = general SC; ξ_2 = academic SC; ξ_3 = English SC; ξ_4 = mathematics SC). GSC = general SC; ASC = academic SC; ESC = English SC; MSC = mathematics SC; SDQGSC = Self Description Questionnaire (SDQ) General-Self subscale; APIGSC = Affective Perception Inventory (API)

of SC Measurements and Structure

High Track

$$
\begin{array}{cccc}
\xi_1 & \xi_2 & \xi_3 & \xi_4
\end{array}
$$

$$
\begin{bmatrix}
1 & 0 & 0 & 0 \\
\lambda_{21} & 0 & 0 & 0 \\
\lambda_{31} & 0 & 0 & 0 \\
0 & 1 & 0 & 0 \\
0 & \lambda_{52} & 0 & 0 \\
\lambda_{61} & 0 & 1 & 0 \\
0 & 0 & \lambda_{73} & 0 \\
0 & 0 & \lambda_{83} & 0 \\
0 & 0 & 0 & 1 \\
0 & 0 & 0 & \lambda_{10.4} \\
0 & 0 & 0 & \lambda_{11.4}
\end{bmatrix}
$$

$$
\begin{bmatrix}
\phi_{11} & & & \\
\phi_{21} & \phi_{22} & & \\
\phi_{31} & \phi_{32} & \phi_{33} & \\
\phi_{41} & \phi_{42} & \phi_{43} & \phi_{44}
\end{bmatrix}
$$

$$
\begin{bmatrix}
\delta_{11} & & & & & & & & & & \\
0 & \delta_{22} & & & & & & & & & \\
0 & 0 & \delta_{33} & & & & & & & & \\
0 & 0 & 0 & \delta_{44} & & & & & & & \\
0 & 0 & 0 & 0 & \delta_{55} & & & & & & \\
0 & 0 & 0 & 0 & 0 & \delta_{66} & & & & & \\
0 & 0 & 0 & 0 & 0 & 0 & \delta_{77} & & & & \\
0 & 0 & 0 & 0 & \delta_{85} & 0 & 0 & \delta_{88} & & & \\
0 & 0 & 0 & 0 & 0 & 0 & 0 & 0 & \delta_{99} & & \\
0 & 0 & 0 & 0 & 0 & 0 & \delta_{10.7} & 0 & 0 & \delta_{10.10} & \\
0 & 0 & 0 & 0 & \delta_{11.5} & 0 & 0 & \delta_{11.8} & 0 & 0 & \delta_{11.11}
\end{bmatrix}
$$

Self-concept subscale; SESGSC = Self-esteem Scale; SDQASC = SDQ Academic SC subscale; SCAASC = Self-concept of Ability Scale (SCA); SDQESC = SDQ Verbal SC subscale; APIESC = API English Perceptions subscale; SCAESC = SCA form B (SC of English ability); SDQMSC = SDQ Mathematics SC subscale; APIMSC = API Mathematics Perceptions subscale; SCAMSC = SCA form C (SC of mathematics ability).

TABLE 8.10. Pattern of LISREL Parameters for Testing the Invariance

Low Track

Λ_Y (factor loadings):

Y	η_1	η_2	η_3	η_4	υ
SDQGSC	1	0	0	0	λ_{15}
APIGSC	λ_{21}	0	0	0	λ_{25}
SESGSC	λ_{31}	0	0	0	λ_{35}
SDQASC	0	1	0	0	λ_{45}
SCAASC	0	λ_{52}	0	0	λ_{55}
SDQESC	0	0	1	0	λ_{65}
APIESC	λ_{71}	0	λ_{73}	0	λ_{75}
SCAESC	0	0	λ_{83}	0	λ_{85}
SDQMSC	0	0	0	1	λ_{95}
APIMSC	0	0	0	$\lambda_{10.4}$	$\lambda_{10.5}$
SCAMSC	0	0	0	$\lambda_{11.4}$	$\lambda_{11.5}$

Ψ (factor variance-covariance matrix):

GSC	ζ_{11}				
ASC	ζ_{21}	ζ_{22}			
ESC	ζ_{31}	ζ_{32}	ζ_{33}		
MSC	ζ_{41}	ζ_{42}	ζ_{43}	ζ_{44}	
DUMMY	0	0	0	0	0

Θ_ϵ (error variance-covariance matrix):

SDQGSC	ϵ_{11}										
APIGSC	0	ϵ_{22}									
SESGSC	0	0	ϵ_{33}								
SDQASC	0	0	0	ϵ_{44}							
SCAASC	0	0	0	0	ϵ_{55}						
SDQESC	0	0	0	0	0	ϵ_{66}					
APIESC	0	0	0	0	0	0	ϵ_{77}				
SCAESC	0	0	0	0	ϵ_{85}	0	0	ϵ_{88}			
SDQMSC	0	0	0	0	0	ϵ_{96}	0	0	ϵ_{99}		
APIMSC	0	0	0	0	0	0	$\epsilon_{10.7}$	0	0	$\epsilon_{10.10}$	
SCAMSC	0	0	0	0	$\epsilon_{11.5}$	0	0	0	0	0	$\epsilon_{11.11}$

Γ (mean estimate vector):

GSC	0
ASC	0
ESC	0
MSC	0
DUMMY	1

Λ_Y = factor loading matrix; Ψ = factor variance-covariance matrix; Θ_ϵ = error variance-covariance matrix; Γ = mean estimate vector; Y = observed measures of self-concept (SC); η_1–η_4 = SC factors (η_1 = general SC; η_2 = academic SC; η_3 = English SC; η_4 = mathematics SC); υ = mean intercepts; GSC = general SC; ASC = academic SC; ESC = English SC; MSC = mathematics SC; SDQGSC = Self Description Questionnaire (SDQ) General-self subscale; APIGSC = Affective Perception Inventory (API) Self concept subscale; SESGSC = Self-esteem Scale; SDQASC = SDQ Academic SC subscale; SCAASC = Self-concept of Ability Scale (SCA); SDQESC = SDQ English SC subscale;

of Mean Structures

High Track

$$
\begin{array}{ccccc}
\eta_1 & \eta_2 & \eta_3 & \eta_4 & \upsilon
\end{array}
$$

$$
\begin{bmatrix}
1 & 0 & 0 & 0 & \lambda_{21} \\
\lambda_{21} & 0 & 0 & 0 & \lambda_{25} \\
\lambda_{31} & 0 & 0 & 0 & \lambda_{35} \\
0 & 1 & 0 & 0 & \lambda_{45} \\
0 & \lambda_{52} & 0 & 0 & \lambda_{55} \\
\lambda_{61} & 0 & 1 & 0 & \lambda_{65} \\
0 & 0 & \lambda_{73} & 0 & \lambda_{75} \\
0 & 0 & \lambda_{83} & 0 & \lambda_{85} \\
0 & 0 & 0 & 1 & \lambda_{95} \\
0 & 0 & 0 & \lambda_{10.4} & \lambda_{10.5} \\
0 & 0 & 0 & \lambda_{11.4} & \lambda_{11.5}
\end{bmatrix}
$$

$$
\begin{bmatrix}
\zeta_{11} & & & & \\
\zeta_{21} & \zeta_{22} & & & \\
\zeta_{31} & \zeta_{32} & \zeta_{33} & & \\
\zeta_{41} & \zeta_{42} & \zeta_{43} & \zeta_{44} & \\
0 & 0 & 0 & 0 & 0
\end{bmatrix}
$$

$$
\begin{bmatrix}
\epsilon_{11} & & & & & & & & & & \\
0 & \epsilon_{22} & & & & & & & & & \\
0 & 0 & \epsilon_{33} & & & & & & & & \\
0 & 0 & 0 & \epsilon_{44} & & & & & & & \\
0 & 0 & 0 & 0 & \epsilon_{55} & & & & & & \\
0 & 0 & 0 & 0 & 0 & \epsilon_{66} & & & & & \\
0 & 0 & 0 & 0 & 0 & 0 & \epsilon_{77} & & & & \\
0 & 0 & 0 & 0 & \epsilon_{85} & 0 & 0 & \epsilon_{88} & & & \\
0 & 0 & 0 & 0 & 0 & 0 & 0 & 0 & \epsilon_{99} & & \\
0 & 0 & 0 & 0 & 0 & 0 & \epsilon_{10.7} & 0 & 0 & \epsilon_{10.10} & \\
0 & 0 & 0 & 0 & \epsilon_{11.5} & 0 & 0 & \epsilon_{11.8} & 0 & 0 & \epsilon_{11.1}
\end{bmatrix}
$$

$$
\begin{bmatrix}
\gamma_{11} \\
\gamma_{21} \\
\gamma_{31} \\
\gamma_{41} \\
1
\end{bmatrix}
$$

APIESC = API English Perceptions subscale; SCAESC = SCA form B (SC of English ability); SDQMSC = SDQ Mathematics SC subscale; APIMSC = Mathematics Perceptions subscale; SCAMSC = SCA form C (SC of mathematics ability).

From Byrne, Shavelson, and Muthén (1989) "Testing for the Equivalence of Factor Covariance and Mean Structures: The Issue of Partial Measurement Invariance" in *Psychological Bulletin, 105*, 456–466. Copyright 1989 by American Psychological Association. Reprinted with permission.

variable, a row of 0s (one for each variable was added to the last row of the input matrix which, in the case of the present data, is a correlation matrix (see Table 8.1); the value of 1.00 was added to the series of standard deviations (i.e., the standard deviation value representing the dummy variable). Fourth, since the analysis of structured means must be based on the moment, rather than on the covariance matrix, the observed mean values were added to the data input; a value of 1.00 was added for the dummy variable since its value was fixed. Fifth, the Λ and Ψ matrices must be modified to accommodate the dummy variable; this was accomplished as follows: and extra column of free λs was added to the matrix; these represent the measurement intercepts and an extra row of 0s was added to the Ψ matrix; ζ_{55} was fixed to zero.[4] Sixth, the latent mean values were estimated in the gamma (Γ) matrix. The parameters γ_{11} to γ_{41} were fixed to zero for the low track, but allowed to be freely estimated for the high track; γ_{51} was fixed to 1.0 for both tracks. Finally, since the analysis of mean structures is based on the moment matrix, the DA card was modified to read as MA = MM.

The LISREL specification input for both the low and high tracks is presented in Table 8.11. To get a clear picture of this transformation from the all-X to the all-Y model, I urge you to study and compare carefully the pattern of parameters presented in Tables 8.9 (all-X model) and 8.10 (all-Y model), and the LISREL specification input in Tables 8.2 (all-X model) and 8.11 (all-Y model).

Since the origins of the measurements and the means of the latent variables cannot be identified simultaneously, absolute mean estimates are not possible. However, when the parameter specifications as described earlier are imposed, latent mean differences between groups can be estimated; one group is used as the reference group and as such, its latent mean parameters are fixed to 0.0. In this case, the low track served as the reference group; mean parameters for the high track were freely estimated. Comparison of the groups, then, is based on the difference from zero. Statistical significance is determined from the T-values (mean estimates divided by their standard error estimates).

Once again, I want to draw your attention to the fact that only the factor-loading parameters known to be consistent in their SC measurements across track were held invariant. Thus, it is important to note that since λ_{71} and λ_{61} were freely estimated for the low and high track, respectively, the intercept terms for these parameters (λ_{75}, λ_{65}) were also free to vary for each track.

---

[4]The LISREL program will print the message that "PSI is not positive definite." This can be ignored since it is a function of ζ_{55} being fixed to 0.0.

TABLE 8.11. LISREL Specification Input for Testing the Invariance of Mean Structures

```
TESTING INV OF MEANS ACROSS TRACK - GRP1= HI TRACK "AGMEANS" FILE
DA NG=2 NI=13 NO=582 MA=MM
LA
*
'SDQGSC' 'SDQASC' 'SDQESC' 'SDQMSC' 'APIGSC' 'SESGSC' 'APIASC' 'SCAASC'
'APIESC' 'SCAESC' 'APIMSC' 'SCAMSC' 'DUMMY'
KM SY
(13F4.3)
1000
 3301000
 267 3971000
 173 427-0111000
 658 323 212 2001000
 812 325 290 226 6681000
 556 624 338 325 618 5671000
 250 660 342 500 266 312 5391000
 151 412 723-040 188 201 408 3411000
 100 415 559-007 107 140 329 536 6941000
 180 455 041 892 272 275 405 516 066 0411000
 133 401 015 843 193 189 351 612-016 086 8241000
 000 000 000 000 000 000 000 000 000 000 000 0000000
ME
*
75.792 57.830 57.569 49.043 76.768 31.467 73.802 30.301 61.794 28.933
47.223 26.223 1.000
SD
*
14.563 11.723 9.867 16.951 9.394 5.063 9.556 4.919 11.191 5.727 11.606
7.986 1.000
SE
1 5 6 2 8 3 9 10 4 11 12 13/
MO NY=11 FI NX=1 NE=5 LY=FU GA=FU,FI PS=SY,FI BE=ZE TE=SY,FI
FR LY(2,1) LY(3,1) LY(5,2) LY(7,3) LY(8,3) LY(10,4) LY(11,4)
FR LY(1,5) LY(2,5) LY(3,5) LY(4,5) LY(5,5) LY(6,5) LY(7,5) LY(8,5)
FR LY(9,5) LY(10,5) LY(11,5)
FR LY(6,1)
FR TE(1,1) TE(2,2) TE(3,3) TE(4,4) TE(5,5) TE(6,6) TE(7,7) TE(8,8)
FR TE(9,9) TE(10,10) TE(11,11)
FR TE(8,5) TE(11,5) TE(10,7) TE(11,8)
FR GA(1,1) GA(2,1) GA(3,1) GA(4,1)
FR PS(1,1) PS(2,2) PS(3,3) PS(4,4)
FR PS(2,1) PS(3,1) PS(4,1) PS(3,2) PS(4,2) PS(4,3)
ST 30.0 LY(1,1) LY(4,2) LY(6,3) LY(9,4)
ST 5.0 LY(6,1)
ST 15.0 LY(2,1) LY(3,1) LY(5,2) LY(7,3) LY(8,3) LY(10,4) LY(11,4)
ST 5.0 LY(1,5) LY(2,5) LY(3,5) LY(4,5) LY(5,5) LY(6,5) LY(7,5) LY(8,5)
ST 5.0 LY(9,5) LY(10,5) LY(11,5)
ST .1 PS(1,1) PS(2,2) PS(3,3) PS(4,4)
ST .02 PS(2,1) PS(3,1) PS(4,1) PS(3,2) PS(4,2) PS(4,3)
ST 5.0 GA(1,1) GA(2,1) GA(3,1) GA(4,1)
ST 1.0 GA(5,1)
ST 40.0 TE(1,1) TE(2,2) TE(3,3) TE(4,4) TE(5,5) TE(6,6) TE(7,7) TE(8,8)
ST 40.0 TE(9,9) TE(10,10) TE(11,11)
ST 6.0 TE(8,5) TE(11,5) TE(10,7) TE(11,8)
OU NS SE TV MI

TESTING FOR MEANS - GROUP 2 - GLEVEL
DA NO=248
LA
*
'SDQGSC' 'SDQASC' 'SDQESC' 'SDQMSC' 'APIGSC' 'SESGSC' 'APIASC' 'SCAASC'
'APIESC' 'SCAESC' 'APIMSC' 'SCAMSC' 'DUMMY'
KM SY
(13F4.3)
1000
 3201000
 307 2981000
 244 355-0551000
 614 237 214 2671000
 755 261 276 255 5881000
 456 571 392 345 547 4581000
 270 580 265 226 219 270 5231000
 143 430 623 030 181 108 476 3731000
 231 377 433 004 265 245 424 509 4981000
 250 388 063 779 245 234 409 345 232 0271000
 234 348-012 719 199 214 362 442 075 077 7421000
 000 000 000 000 000 000 000 000 000 000 000 0000000
ME
```

TABLE 8.11. Continued

```
☆
75.936 49.415 55.036 41.569 76.758 31.157 70.165 24.746 57.794 25.343
41.710 22.944 1.000
SD
☆
13.442 12.391 9.468 13.416 9.028 4.875 8.830 4.480 10.701 4.858 10.566
5.824 0.000
SE
1  5  6  2  8  3  9  10  4  11  12  13/
MO LY=FU PS=SY,FI TE=SY,FI
FR LY(8,3) LY(10,4)
FR LY(6,5) LY(7,5)
FR LY(7,1)
FR PS(1,1) PS(2,2) PS(3,3) PS(4,4)
FR PS(2,1) PS(3,1) PS(4,1) PS(3,2) PS(4,2) PS(4,3)
FR TE(1,1) TE(2,2) TE(3,3) TE(4,4) TE(5,5) TE(6,6) TE(7,7) TE(8,8)
FR TE(9,9) TE(10,10) TE(11,11)
FR TE(8,5) TE(11,5) TE(10,7) TE(9,6)
VA 1.0 GA(5,1)
ST 30.0 LY(1,1) LY(4,2) LY(6,3) LY(9,4)
ST 5.0 LY(8,3) LY(10,4) LY(6,5) LY(7,5)
ST 5.0 LY(7,1)
ST .1 PS(1,1) PS(2,2) PS(3,3) PS(4,4)
ST .02 PS(2,1) PS(3,1) PS(4,1) PS(3,2) PS(4,2) PS(4,3)
ST 40.0 TE(1,1) TE(2,2) TE(3,3) TE(4,4) TE(5,5) TE(6,6) TE(7,7) TE(8,8)
ST 40.0 TE(9,9) TE(10,10) TE(11,11)
ST 6.0 TE(8,5) TE(11,5) TE(10,7) TE(9,6)
EQ LY(1,2,1) LY(2,1)
EQ LY(1,3,1) LY(3,1)
EQ LY(1,5,2) LY(5,2)
EQ LY(1,7,3) LY(7,3)
EQ LY(1,11,4) LY(11,4)
EQ LY(1,1,5) LY(1,5)
EQ LY(1,2,5) LY(2,5)
EQ LY(1,3,5) LY(3,5)
EQ LY(1,4,5) LY(4,5)
EQ LY(1,5,5) LY(5,5)
EQ LY(1,8,5) LY(8,5)
EQ LY(1,9,5) LY(9,5)
EQ LY(1,10,5) LY(10,5)
EQ LY(1,11,5) LY(11,5)
OU NS SE
```

3.2. LISREL Output

The parameter estimates and standard errors are shown in Table 8.12. At first glance, the format used in presenting the results may seem somewhat confusing to you; your reaction is not unexpected. Admittedly, the presentation of these data, although conventional, certainly bears further explanation. Let's now examine important elements in the table.

1. All values in parentheses, as footnoted in Table 8.12, represent the standard errors of the estimates.

2. All estimates appearing in the center column represent parameters that were held invariant across track; the values presented are therefore common to both tracks. All other values relate to parameters that were freely estimated for each group.[5] In this regard, let us now look more closely at particular measurement parameters that were specified as noninvariant across track. These include:

---

[5]Except for parameters γ_{11}–γ_{41} for the low track; these values were fixed to 0.0.

TABLE 8.12. Maximum Likelihood Estimates for Self-Concept, Facets[a]

Parameter	Low track	High track
$\nu_1(\lambda_{15})$		75.71 (0.81)
$\nu_2(\lambda_{25})$		76.69 (0.47)
$\nu_3(\lambda_{35})$		31.34 (0.29)
$\nu_4(\lambda_{45})$		47.55 (0.67)
$\nu_5(\lambda_{55})$		25.20 (0.28)
$\nu_6(\lambda_{65})$	55.07 (0.61)	52.35 (0.69)
$\nu_7(\lambda_{75})$	58.03 (0.65)	54.36 (0.88)
$\nu_8(\lambda_{85})$		25.61 (0.31)
$\nu_9(\lambda_{95})$		41.72 (0.81)
$\nu_{10}(\lambda_{10.5})$		42.17 (0.58)
$\nu_{11}(\lambda_{11.5})$		23.05 (0.35)
λ_{21}		16.00 (0.66)
λ_{31}		10.69 (0.35)
λ_{52}		13.72 (0.56)
λ_{73}		42.71 (0.94)
λ_{83}	13.71 (1.31)	18.53 (0.94)
$\lambda_{10.4}$	23.98 (1.21)	20.26 (0.43)
$\lambda_{11.4}$		12.83 (0.31)
λ_{61}		3.56 (0.69)
λ_{71}	−6.98 (1.66)	
$\theta\epsilon_{11}$	40.03 (6.81)	44.72 (4.86)
$\theta\epsilon_{22}$	42.06 (4.29)	40.33 (2.69)
$\theta\epsilon_{33}$	6.38 (0.94)	4.25 (0.57)
$\theta\epsilon_{44}$	77.80 (8.72)	52.95 (4.16)
$\theta\epsilon_{55}$	8.88 (1.34)	8.05 (0.78)
$\theta\epsilon_{66}$	42.53 (4.76)	37.12 (2.73)
$\theta\epsilon_{77}$	28.16 (6.30)	17.67 (3.38)
$\theta\epsilon_{88}$	14.43 (1.48)	14.34 (1.05)
$\theta\epsilon_{99}$	41.86 (5.72)	25.91 (3.22)
$\theta\epsilon_{10.10}$	22.79 (3.65)	16.56 (1.65)
$\theta\epsilon_{11.11}$	10.92 (1.25)	14.30 (1.00)
$\theta\epsilon_{85}$	4.67 (0.96)	6.12 (0.66)
$\theta\epsilon_{10.7}$	17.60 (3.17)	7.65 (1.41)
$\theta\epsilon_{11.5}$	3.70 (0.85)	5.69 (0.65)
$\theta\epsilon_{96}$	−8.37 (3.63)	
$\theta\epsilon_{11.8}$		3.20 (0.72)
$\gamma_{11}(GSC)$	0.0	0.01 (0.03)
$\gamma_{21}(ASC)$	0.0	0.36 (0.03)
$\gamma_{31}(ESC)$	0.0	0.17 (0.02)
$\gamma_{41}(MSC)$	0.0	0.25 (0.03)
ζ_{11}	0.15 (0.12)	0.19 (0.01)
ζ_{22}	0.07 (0.01)	0.09 (0.01)
ζ_{33}	0.05 (0.01)	0.06 (0.01)
ζ_{44}	0.15 (0.02)	0.29 (0.02)

[a]Standard errors are in parentheses. $\chi^2_{(176)} = 201.82$ GSC = general self-concept (SC); ASC = academic SC; ESC = English SC; MSC = mathematics SC.

From Byrne, Shavelson, and Muthén (1989) "Testing for the Equivalence of Factor Covariance and Mean Structures: The Issue of Partial Measurement Invariance" in *Psychological Bulletin, 105,* 456–466. Copyright 1989 by American Psychological Association. Reprinted with permission.

(a) $v_6(\lambda_{65})$ and $v_7(\lambda_{75})$—the intercept terms for the two cross-loadings (λ_{61}; λ_{71}).

(b) λ_{83} and $\lambda_{10,4}$—the measures of English SC (SCAESC) and mathematics SC (APIMSC).

(c) λ_{61} and λ_{71}—the cross-loading of two English SC subscales onto the general SC factor for the high track (λ_{61}; SDQESC) and for the low track (λ_{71}; APIESC), respectively. Thus, only one parameter for each group is estimated.

(d) $\theta_{\epsilon96}$ and $\theta_{\epsilon11,8}$—error covariances specific to the high track (SDQMSC/SDQESC) and low track (SCAMSC/SCAESC), respectively.[6]

(e) $\theta_{\epsilon11}$–$\theta_{\epsilon11,11}$—error variances specific to each track.

3. In examining the latent mean parameters (γ_{11} to γ_{41}), we see that estimates are presented only for the high track; recall that these parameters were specified as fixed to 0.0 for the low track. For interpretative purposes, the following points are worthy of note:

(a) The fact that the γ estimates for the high track are positive values indicates that scores for all SC factors were higher for the high track than for the low track.

(b) The largest differences between tracks are shown to be in academic SC (γ_{21}), followed by mathematics SC (γ_{41}) and English SC (γ_{31}), respectively; mean track differences in general SC (γ_{11}) were negligible.

(c) To determine if the differences in latent SC means are statistically significant, we examine the T-values presented in the output. Since the parameters for the low track are fixed to 0.0, the T-values for the high track indicate whether the estimates are significantly different from zero; values >2.00 are considered statistically significant. As such, significant mean track differences were found with respect to academic SC ($T = 13.99$), English SC ($T = 7.58$), and mathematics SC ($T = 6.93$); differences in general SC were not significant ($T = 0.23$).

Overall, the results demonstrate that the test for invariant SCs across track based on mean and covariance structures was statistically more powerful than tests based on covariance structures alone. Whereas tests of invariance based on the latter found academic track differences in mathematics SC (ϕ_{44}) only, this was not so in the analysis that also included mean structures; significant differences were also found in academic and English SCs.

---

[6]Note that while the error covariances $\theta_{\epsilon85}$, $\theta_{\epsilon10,7}$, and $\theta_{\epsilon11,5}$ were common to each track, they were not constrained equal across groups.

4. Summary

This chapter demonstrated applications of two unique concepts associated with tests for invariance: testing for and with partial measurement invariance and testing for differences in latent mean structures. Since the LISREL approach to testing for mean structures requires total reparameterization if the analysis of covariance structures was based on an all-X model, the transposition of parameters from an all-X to an all-Y specification was detailed. Relatedly, the creation and specification of the dummy variable in LISREL analyses of mean structures was explained and demonstrated. Finally, results were interpreted within the tabular framework typically used in reporting findings from the analysis of mean structures.

References

Alwin, D.F., & Jackson, D.G. (1980). Measurement models for response errors in surveys: Issues and applications. In K.F. Schuessler (Ed.), *Sociological methodology* (pp. 68–119). San Francisco: Jossey-Bass.

Anderson, J.C., & Gerbing, D.W. (1988). Structural equation modeling in practice: A review and recommended two-step approach. *Psychological Bulletin, 103*, 411–423.

Benson, J. (1987). Detecting item bias in affective scales. *Educational and Psychological Measurement, 47*, 55–67.

Bentler, P.M. (1985). Theory and implementation of EQS: A structural equations program. Los Angeles: BMDP Statistical Software.

Bentler, P.M., & Bonett, D.G. (1980). Significance tests and goodness-of-fit in the analysis of covariance structures. *Psychological Bulletin, 88*, 588–606.

Boomsma, A. (1982). The robustness of LISREL against small sample sizes in factor analysis models. In H. Wold and K. Joreskog (Eds.), *Systems under indirect observation* (pp. 149–173). New York: Elsevier North-Holland.

Brookover, W.B. (1962). *Self-concept of Ability Scale.* East Lansing, Mich.: Educational Publication Services.

Byrne, B.M. (1983). Investigating measures of self-concept. *Measurement and Evaluation in Guidance, 16*, 115–126.

Byrne, B.M. (1988a). Adolescent self-concept, ability grouping, and social comparison: Reexamining academic track differences in high school. *Youth and Society, 20*, 46–67.

Byrne, B.M. (1988b). Measuring adolescent self-concept: Factorial validity and equivalency of the SDQ III across gender. *Multivariate Behavioral Research, 24*, 361–375.

Byrne, B.M. (1988c). The Self Description Questionnaire III: Testing for equivalent factorial validity across ability. *Educational and Psychological Measurement, 48*, 397–406.

Byrne, B.M. (in press). Multigroup comparisons and the assumption of equivalent construct validity across groups: Methodological and substantive issues. *Multivariate Behavioral Research.*

Byrne, B.M., & Shavelson, R.J. (1986). On the structure of adolescent self-concept. *Journal of Educational Psychology, 78*, 474–481.

Byrne, B.M., & Shavelson, R.J. (1987). Adolescent self-concept: Testing the assumption of equivalent structure across gender. *American Educational Research Journal, 24,* 365–385.

Byrne, B.M., Shavelson, R.J., & Muthén, B. (1989). Testing for the equivalence of factor covariance and mean structures: The issue of partial measurement invariance. *Psychological Bulletin, 105,* 456–466.

Campbell, D.T., & Fiske, D.W. (1959). Convergent and discriminant validation by the mulitrait-multimethod matrix. *Psychological Bulletin, 56,* 81–105.

Carmines, E.G., & McIver, S.P. (1981). Analyzing models with unobserved variables: Analysis of covariance structures. In G.W. Bohrnstedt and E.F. Borgatta (Eds.), *Social measurement: Current issues* (pp. 65–115). Beverly Hills, Calif.: Sage.

Cliff, N. (1983). Some cautions concerning the application of causal modeling methods. *Multivariate Behavioral Research, 18,* 115–126.

Cole, D.A., & Maxwell, S.E. (1985). Multitrait-multimethod comparisons across populations: A confirmatory factor analytic approach. *Multivariate Behavioral Research, 20,* 389–417.

Cudeck, R., & Browne, M.W. (1983). Cross-validation of covariance structures. *Multivariate Behavioral Research, 18,* 147–167.

Gerbing, D.W., & Anderson, J.C. (1984). On the meaning of within-factor correlated measurement errors. *Journal of Consumer Research, 11,* 572–580.

Green, D.R. (1975). What does it mean to say a test is biased? *Education and Urban Society, 8,* 33–52.

Jöreskog, K.G. (1971a). Simultaneous factor analysis in several populations. *Psychometrika, 36,* 409–426.

Jöreskog, K.G. (1971b). Statistical analysis of sets of congeneric tests. *Psychometrika, 36,* 109–133.

Jöreskog, K.G. (1983). UK LISREL Workshop, University of Edinburgh, Scotland.

Jöreskog, K.G., & Sorböm, D. (1985). *LISREL VI: Analysis of linear structural relationships by the method of maximum likelihood.* Mooresville, Ind.: Scientific Software Inc.

Jöreskog, K.G., & Sorböm, D. (1986). *PRELIS A program for multivariate data screening and data summarization: A preprocessor for LISREL.* Mooresville, Ind.: Scientific Software Inc.

Long, J. Scott (1983). *Confirmatory factor analysis.* Beverly Hills, Calif.: Sage.

MacCallum, R. (1986). Specification searches in covariance structure modeling. *Psychological Bulletin, 100,* 107–120.

Marsh, H.W., Balla, J.R., & McDonald, R.P. (1988). Goodness-of-fit indexes in confirmatory factor analysis: The effect of sample size. *Psychological Bulletin, 103,* 391–410.

Marsh, H.W., Byrne, B.M., & Shavelson, R.J. (1988). A multifaceted academic self-concept: Its hierarchical structure and its relation to academic achievement. *Journal of Educational Psychology, 80,* 366–380.

Marsh, H.W., & Hocevar, D. (1983). Confirmatory factory analysis of multitrait-multimethod matrices. *Journal of Educational Measurement, 20,* 231–248.

Marsh, H.W., & Hocevar, D. (1985). Application of confirmatory factor analysis to the study of self-concept: First- and higher order factor models and their invariance across groups. *Psychological Bulletin, 97*(3), 562–582.

Marsh, H.W., & O'Neill, R. (1984). Self Description Questionnaire III: The construct validity of multidimensional self-concept ratings by late adolescents. *Journal of Educational Measurement, 21,* 153–174.

McDonald, R.P. (1978). A simple comprehension model for the analysis of covariance structures. *British Journal of Mathematical and Statistical Psychology, 31,* 59–72.

McGaw, B., & Joreskog, K.G. (1971). Factorial invariance of ability measures in groups differing in intelligence and socio-economic status. *British Journal of Mathematical and Statistical Psychology, 24,* 154–168.

Muthén, B. (1987). *LISCOMP: Analysis of linear structural relations using a comprehension measurement model.* Mooresville, Ind.: Scientific Software Inc.

Rock, D.A., Werts, C.E., & Flaugher, R.L. (1978). The use of analysis of covariance structures for comparing the psychometric properties of multiple variables across populations. *Multivariate Behavioral Research, 13,* 403–418.

Rosenberg, M. (1965). *Society and the adolescent self-image.* Princeton, N.J.: Princeton University Press.

Saris, W., & Stronkhorst, H. (1984). *Causal modelling in nonexperimental research.* Amsterdam: Sociometric Research Foundation.

Schmitt, N., & Stults, D.M. (1986). Methodology review: Analysis of multitrait-multimethod matrices. *Applied Psychological Measurement, 10,* 1–22.

Shavelson, R.J., Hubner, J.J., & Stanton, G.C. (1976). Self-concept: Validation of construct interpretations. *Review of Educational Research, 46* 407–441.

Shavelson, R.J., & Stuart, K.R. (1981). Application of causal modeling to the validation of self-concept interpretation of test scores. In M.D. Lynch, K. Gregan, and A.A. Norem-Hebelson (Eds.), *Self-concept: Advances in theory and research* (pp. 223–235). Cambridge, Mass.: Ballinger.

Soares, A.T., & Soares, L.M. (1979). *The Affective Perception Inventory—Advanced level.* Trumbell, Conn.: ALSO.

Soares, A.T., & Soares, L.M. (1980). *The Affective Perception Inventory: Test manual/advanced level.* Trumbell, Conn.: ALSO.

Tanaka, J.S., & Huba, G.J. (1984). Confirmatory hierarchical factor analyses of psychological distress measures. *Journal of Personality and Social Psychology, 46,* 621–635.

Tanaka, J.S., & Huba, G.J. (1985). A fit index for covariance structure models under arbitrary GLS estimation. *British Journal of Mathematical and Statistical Psychology, 38,* 197–201.

Werts, C.E., Rock, D.A., Linn, R.L., & Joreskog, K.G. (1976). Comparison of correlations, variances, covariances, and regression weights with or without measurement error. *Psychological Bulletin, 83,* 1007–1013.

Wheaton, B., Muthén, B., Alwin, D.F., & Summers, G.F. (1977). Assessing reliability and stability in panel models. In D.R. Heise (Ed.), *Sociological methodology, 1977* (pp. 84–136). San Francisco: Jossey-Bass.

Widaman, K.F. (1985). Hierarchically tested covariance structure models for multitrait-multimethod data. *Applied Psychological Measurement, 9,* 1–26.

Appendix: Description of Data and Measuring Instruments

Sample and Procedure

The original sample comprised 996 grade 11 and 12 students from two suburban high schools in Ottawa, Canada. The data approximated a normal distribution with skewness ranging from -1.27 to 0.06 ($\bar{X} = -0.32$); kurtosis ranged from -0.86 to 2.30 ($\bar{X} = 0.12$).

A battery of SC instruments (described later) was administered to intact classroom groups during one 50- minute period. To ensure the relevancy of self-concept (SC) responses related to English and mathematics, it was important that all students in the sample be enrolled in both of these subject areas. Since English is part of the core curriculum for high schools in Ontario (i.e., compulsory), it was known that all students were enrolled in at least one English course; therefore, only mathematics classes were tested for the study. The testing was completed approximately two weeks following the April report cards. The students therefore had the opportunity of being fully cognizant of their academic performance prior to completing the tests for the study. This factor was considered important in the measurement of academic and subject-specific SCs.

Instrumentation

The SC test battery consisted of 12 measures—3 each for general SC, academic SC, English SC, and mathematics SC. All were self-report rating scales designed for use with a high school population. General SC was measured using the General-self Subscale of the Self Description Questionnaire (SDQIII) (Marsh & O'Neill, 1984), the Self-concept subscale of the Affective Perception Inventory (API) (Soares & Soares, 1979), and the Self-esteem Scale (SES) (Rosenberg, 1965). Measures of academic SC were the SDQIII Academic Self-concept scale, the API Student Self subscale, and the Self-concept of Ability Scale form A (SCA) (Brookover,

1962). English SC was measured with the SDQIII Verbal Self-concept subscale, the API English Perceptions subscale, and the SCA form B. Items on form B are identical to those on form A, except that they elicit responses relative to specific academic content (e.g., "How do you rate you ability in English compared to your close friends'?"). Finally, measures of mathematics SC included the SDQIII Mathematics SC subscale, the API Mathematics Perceptions subscale, and the SCA form C (items specific to mathematics ability). The instruments were selected because they purported to measure (with some empirical justification) the SC facets in the theory to be tested.

The SDQIII is structured on an eight-point Likert-type scale with responses ranging form "1—Definitely False" to "8—Definitely True." The general-self subscale contains 12 items and the other three subscales, 10 items each. Internal consistency reliability coefficients ranging from 0.86 to 0.93 (Md α = 0.90) for the SDQIII general SC and each of the academic SC subscales, and strong support for their construct validity based on interpretations consistent with the Shavelson et al. (1976) model of SC have been reported (Byrne & Shavelson, 1986; Marsh & O'Neill, 1984). These four subscales have also been shown to be invariant across sex and ability (Byrne, 1988b, 1988c).

The API was developed as a semantic differential with a forced-choice format containing four categories spread along a continuum between two dichotomous terms (e.g., "happy," "unhappy"). Internal consistency coefficients ranging from 0.79 to 0.95 (Md α = 0.85) have been reported for the subscale measures of general SC, academic SC, and the subject-specific SCs (Byrne & Shavelson, 1986; Soares & Soares, 1980). Convergent validity coefficients ranged from 0.49 to 0.55 (Md r = 0.50) with peer ratings, and from 0.37 to 0.74 (Md r = 48.5) with teacher ratings for the same subscales. Soares and Soares (1980) also presented evidence of discriminant validity. The number of items comprising each of the API subscales is as follows: Self-concept—25; Student Self—25; English Perceptions—22; Mathematics Perceptions—17.

The SES is a 10-item Guttman scale based on a format ranging from "strongly agree" to "strongly disagree." A test-retest reliability of 0.62 (Byrne, 1983), and an internal consistency reliability coefficient of 0.87 (Byrne & Shavelson, 1986) have been reported, as well as convergent validities ranging from 0.56 to 0.67 (see Byrne, 1983). The eight-item SCA, also a Guttman scale, is based on a five-point format. Respondents are asked to rank their ability in comparison with others on a scale from 1 ("I am the poorest") to 5 ("I am the best"). Test-retest and internal consistency reliability coefficients ranging from 0.69 to 0.72, and from 0.77 to 0.94, respectively, have been reported (see Byrne, 1983; Byrne & Shavelson, 1986).

Index